DEVELOPMENT AND APPLICATION OF OPEN SOURCE MODELLING AND SIMULATION TOOLS FOR NUCLEAR REACTOR ANALYSIS

The following States are Members of the International Atomic Energy Agency:

AFGHANISTAN
ALBANIA
ALGERIA
ANGOLA
ANTIGUA AND BARBUDA
ARGENTINA
ARMENIA
AUSTRALIA
AUSTRIA
AZERBAIJAN
BAHAMAS, THE
BAHRAIN
BANGLADESH
BARBADOS
BELARUS
BELGIUM
BELIZE
BENIN
BOLIVIA, PLURINATIONAL
  STATE OF
BOSNIA AND HERZEGOVINA
BOTSWANA
BRAZIL
BRUNEI DARUSSALAM
BULGARIA
BURKINA FASO
BURUNDI
CABO VERDE
CAMBODIA
CAMEROON
CANADA
CENTRAL AFRICAN
  REPUBLIC
CHAD
CHILE
CHINA
COLOMBIA
COMOROS
CONGO
COOK ISLANDS
COSTA RICA
CÔTE D'IVOIRE
CROATIA
CUBA
CYPRUS
CZECH REPUBLIC
DEMOCRATIC REPUBLIC
  OF THE CONGO
DENMARK
DJIBOUTI
DOMINICA
DOMINICAN REPUBLIC
ECUADOR
EGYPT
EL SALVADOR
ERITREA
ESTONIA
ESWATINI
ETHIOPIA
FIJI
FINLAND
FRANCE
GABON
GAMBIA, THE

GEORGIA
GERMANY
GHANA
GREECE
GRENADA
GUATEMALA
GUINEA
GUYANA
HAITI
HOLY SEE
HONDURAS
HUNGARY
ICELAND
INDIA
INDONESIA
IRAN, ISLAMIC REPUBLIC OF
IRAQ
IRELAND
ISRAEL
ITALY
JAMAICA
JAPAN
JORDAN
KAZAKHSTAN
KENYA
KOREA, REPUBLIC OF
KUWAIT
KYRGYZSTAN
LAO PEOPLE'S DEMOCRATIC
  REPUBLIC
LATVIA
LEBANON
LESOTHO
LIBERIA
LIBYA
LIECHTENSTEIN
LITHUANIA
LUXEMBOURG
MADAGASCAR
MALAWI
MALAYSIA
MALI
MALTA
MARSHALL ISLANDS
MAURITANIA
MAURITIUS
MEXICO
MONACO
MONGOLIA
MONTENEGRO
MOROCCO
MOZAMBIQUE
MYANMAR
NAMIBIA
NEPAL
NETHERLANDS,
  KINGDOM OF THE
NEW ZEALAND
NICARAGUA
NIGER
NIGERIA
NORTH MACEDONIA
NORWAY
OMAN

PAKISTAN
PALAU
PANAMA
PAPUA NEW GUINEA
PARAGUAY
PERU
PHILIPPINES
POLAND
PORTUGAL
QATAR
REPUBLIC OF MOLDOVA
ROMANIA
RUSSIAN FEDERATION
RWANDA
SAINT KITTS AND NEVIS
SAINT LUCIA
SAINT VINCENT AND
  THE GRENADINES
SAMOA
SAN MARINO
SAUDI ARABIA
SENEGAL
SERBIA
SEYCHELLES
SIERRA LEONE
SINGAPORE
SLOVAKIA
SLOVENIA
SOMALIA
SOUTH AFRICA
SPAIN
SRI LANKA
SUDAN
SWEDEN
SWITZERLAND
SYRIAN ARAB REPUBLIC
TAJIKISTAN
THAILAND
TOGO
TONGA
TRINIDAD AND TOBAGO
TUNISIA
TÜRKİYE
TURKMENISTAN
UGANDA
UKRAINE
UNITED ARAB EMIRATES
UNITED KINGDOM OF
  GREAT BRITAIN AND
  NORTHERN IRELAND
UNITED REPUBLIC OF TANZANIA
UNITED STATES OF AMERICA
URUGUAY
UZBEKISTAN
VANUATU
VENEZUELA, BOLIVARIAN
  REPUBLIC OF
VIET NAM
YEMEN
ZAMBIA
ZIMBABWE

The Agency's Statute was approved on 23 October 1956 by the Conference on the Statute of the IAEA held at United Nations Headquarters, New York; it entered into force on 29 July 1957. The Headquarters of the Agency are situated in Vienna. Its principal objective is "to accelerate and enlarge the contribution of atomic energy to peace, health and prosperity throughout the world".

TECHNICAL REPORTS SERIES No. 496

DEVELOPMENT AND APPLICATION OF OPEN SOURCE MODELLING AND SIMULATION TOOLS FOR NUCLEAR REACTOR ANALYSIS

INTERNATIONAL ATOMIC ENERGY AGENCY
VIENNA, 2025

© IAEA, Copyright Year

Printed by the IAEA in Austria

December 2025

STI/DOC/010/496

https://doi.org/10.61092/iaea.tokn-e953

IAEA Library Cataloguing in Publication Data

Names: International Atomic Energy Agency.
Title: Development and application of open source modelling and simulation tools for nuclear reactor analysis / International Atomic Energy Agency.
Description: Vienna : International Atomic Energy Agency, 2025. | Series: Technical reports series (International Atomic Energy Agency), ISSN 0074-1914 ; no. 496 | Includes bibliographical references.
Identifiers: IAEAL 25-01798 | ISBN 978-92-0-101425-2 (paperback : alk. paper) | ISBN 978-92-0-101525-9 (pdf) | ISBN 978-92-0-101625-6 (epub)
Subjects: LCSH: Nuclear reactors — Design and construction. | Nuclear reactors — Technological innovations. | Nuclear reactors — Training of — Employees. | Nuclear reactors — Analysis.
Classification: UDC 621.039.53 | STI/DOC/010/496

FOREWORD

The IAEA offers its Member States a wide spectrum of research, education and training activities. One such activity is the IAEA-facilitated Open-source Nuclear Codes for Reactor Analysis (ONCORE) initiative, an international collaboration framework focused on the development and implementation of open source multiphysics simulation tools for nuclear reactor analysis. These codes are used to simulate the behaviour of nuclear reactors and predict their performance under different operating conditions. They are a crucial tool for ensuring the safety and reliability of nuclear power plants, as well as for advancing the development of new reactor designs and technologies. ONCORE supports research, education and training for the analysis of advanced nuclear power reactors. Institutions and individuals involved in ONCORE can collaborate and benefit from the development of open source software in the nuclear science and technology field. Open source codes and tools are becoming increasingly important for nuclear reactor analysis because of their numerous benefits. They are developed in a transparent manner, with the source code being publicly available for scrutiny and modification. This transparency allows for increased trust in the results obtained from these codes and tools, as users can verify that the algorithms and models used are sound and accurate. Open source codes and tools enable collaboration and innovation among researchers and developers, as they can freely share their code and build upon the work of others. This results in faster progress and the development of more advanced and efficient tools for nuclear reactor analysis. The use of open source codes and tools can save significant costs associated with licensing closed source software. This is particularly important for smaller organizations or developing countries that may not have the resources to invest in expensive commercial software. Open source codes and tools provide an excellent platform for education and training in nuclear reactor analysis. They can be freely accessed and used by students, researchers and industry professionals, allowing the development of a skilled workforce that is well versed in the latest techniques and technologies.

This publication provides an overview of the benefits and latest developments in the field of open source tools, as well as a detailed overview of available open source tools for reactor analysis and related topics. It also explores the challenges and opportunities associated with the use of open source codes in the nuclear industry. This publication is based on the papers presented during the IAEA Technical Meeting on the Development and Application of Open-Source Modelling and Simulation Tools for Nuclear Reactors, held in Milan, Italy, 20–24 June 2022, as well as a state of the art overview of open source tools from experts in the field.

The IAEA wishes to thank all of the participants of the technical meeting and contributors to this publication for their efforts, in particular C. Fiorina (United States of America). The IAEA officers responsible for this publication were V. Kriventsev and N. Morelová of the Division of Nuclear Power.

ACKNOWLEDGEMENTS

The IAEA gratefully acknowledges the contributions of A. Davis (United Kingdom), C. Demazière (Sweden), B. Forget (United States of America), A. Hébert (Canada), K. Ivanov (United States of America), S. Kelm (Germany), M. Munk (United States of America), P. Romano (United States of America), E. Shwageraus (United Kingdom), N. Touran (United States of America) and P. Wilson (United States of America).

EDITORIAL NOTE

CONTENTS

1. INTRODUCTION

1.1. BACKGROUND

Many institutions have invested several years of effort into developing modern open source tools for analysing nuclear reactors, and the scientific community has embraced the shift towards open source codes and open access data. Recognizing this trend, and to streamline and support this development, the IAEA established the Open-source Nuclear Codes for Reactor Analysis (ONCORE) initiative. The resulting platform is accessible to all Member States, making it a worldwide hub for developers seeking collaboration and for Member States exploring new nuclear codes for research, development, education and training. This platform is particularly valuable for Member States developing nuclear energy capabilities and embarking countries with limited resources for commercial software.

The development of building codes using publicly available numerical libraries and code platforms offers distinct advantages, including code flexibility, high numerical performance tailored to modern computer architectures and the ability to harness in-kind contributions from a large network of institutional partners through the open access philosophy. These features make it an ideal resource for universities and research institutions and contribute to human capacity building in the field through education and training activities.

This publication serves as a comprehensive review of the current state of the field, featuring insights from experts, experiences, the latest innovations and technological challenges in open source software (OSS) and open access data. It also highlights ongoing collaborations supporting research and development (R&D), as well as education and training in nuclear science and technology, drawing on the papers presented during the Technical Meeting on the Development and Application of Open Source Modelling and Simulation Tools for Nuclear Reactors, organized within the framework of the ONCORE initiative and held in June 2022.

1.2. OBJECTIVE

The objectives of this publication are as follows:

— To provide an overview and discussion of the current R&D status of open source nuclear computer codes;
— To discuss the available open source multiphysics simulation tools for nuclear reactor analysis and their state of development;

— To assess the benefits and limitations of the open source model in nuclear software development;
— To address the technical challenges associated with open source tools, including development, maintenance, distribution, the need for open access data, interactions with proprietary codes and export control;
— To explore best practices for OSS projects, including guidelines for initial publication, quality assurance (QA), developer–user interactions, and interoperability with other codes and projects;
— To highlight the educational and training opportunities offered by OSS;
— To identify R&D needs and gaps in order to determine future requirements in the field, focusing efforts on key areas;
— To summarize the work presented by the participants at the June 2022 technical meeting;
— To promote and facilitate the exchange of information within the scope of the technical meeting and its related topics at the national and international levels;
— To support relevant education and training programmes and initiatives in Member States.

Guidance and recommendations provided here in relation to identified good practices represent expert opinion but are not made on the basis of a consensus of all Member States.

1.3. SCOPE

This publication focuses on the current status and progress of development of open source tools, addressing challenges specific to open source projects in the nuclear field, outlining best practices for nuclear open source initiatives, exploring the use of OSS for educational and training purposes, and considering advanced applications of open source tools for solving nuclear engineering challenges. It provides insights into the R&D efforts related to open source codes and tools for nuclear science and technology discussed at the technical meeting.

1.4. STRUCTURE

Section 2 discusses the issues with closed source software (CSS) and provides a historical survey of codes. Section 3 provides a general overview of the currently available tools. Section 4 provides an overview of the differences between open source and proprietary codes, including community building, funding and incentive structures, as well as the adoption of OSS and associated quality, licensing and distribution issues. Section 5 summarizes and discusses the integration process, community management, documentation, intellectual property management and peculiarities of nuclear open source projects. Section 6 focuses on topics such as learning to use a modelling software, investigating physical phenomena using a modelling software and learning the modelling algorithms and modelling skills. Section 7 summarizes the publication, highlighting the gaps and opportunities for collaboration and stating a path forward. The Annex contains summaries of the papers presented at the IAEA Technical Meeting on the Development and Application of Open Source Modelling and Simulation Tools for Nuclear Reactors in June 2022. All papers included in this publication have been peer reviewed and were discussed at the technical meeting.

2. THE OPEN SOURCE APPROACH IN THE NUCLEAR FIELD

2.1. USE OF OPEN SOURCE CODES IN NUCLEAR ENGINEERING

In nuclear science and engineering, the analysis of nuclear reactors plays a pivotal role in ensuring their safe and efficient operation. Accurately modelling the behaviour of nuclear reactors calls for complex simulations and computations. The nuclear reactors at the heart of a range of applications — from electricity generation to medical isotope production and advanced scientific research — involve intricate physical and thermal processes. Understanding their behaviour necessitates sophisticated modelling and simulation techniques, which in turn calls for complex computations. Traditional proprietary software for nuclear reactor analysis poses barriers in terms of cost, accessibility and transparency. Open source codes have emerged as a response to these challenges, providing powerful tools that promote inclusivity and collaboration among researchers and engineers through freely available, collaborative software solutions. Open source codes have already made substantial contributions to nuclear reactor analysis, empowering researchers to study complex phenomena, optimize reactor designs and enhance safety measures. As the nuclear industry continues to evolve, OSS will play an increasingly important role in enabling innovation, collaboration and knowledge dissemination. By fostering a dedicated community of contributors, these codes pave the way for a sustainable and robust future for nuclear reactor analysis.

OSS is typically defined as a computer software whose source is open for the public to access, use, modify and redistribute (e.g. Ref. [1]). The source code is public and can be modified by other users and organizations. CSS refers to computer software whose source code is closed to public access; only the individual or organization that owns the software can change it or authorize someone to change it. CSS is often associated with a licence fee. In CSS, the vendor may be responsible for the results provided by the software. Free software focuses on the freedom that users have with the software, including (i) using the software for any purpose, (ii) having access to the source code, (iii) freely redistributing copies, and (iv) modifying the source code and sharing modifications. Free OSS combines the benefits of both — the user freedom of free software and the open development of OSS.

Open access data are typically defined as data that are free to access, use, reuse and redistribute. Proprietary data means information obtained from a lessee that constitutes trade secrets, or commercial or financial information that is privileged or confidential, or other information that may be withheld under regional law. The term 'proprietary data' is usually associated with data owned by an individual or organization that are deemed important enough to give a competitive advantage to that individual or organization. These data can be protected under copyright laws or patents.

Development of OSS is a trend that was initiated in disciplines not specific to the nuclear reactor domain. Development of OSS is a step towards a more universal outcome: the advent of open science and technology and the advent of a software heritage[1] concept for technology.

The nuclear engineering community is small compared with other engineering fields, and it contains highly specialized, complex and nuanced knowledge. Thus, there is a risk that some knowledge could be lost as people retire. The open source approach provides a medium through which detailed technical knowledge can be expressed and maintained across generations in a structured and operable form.

The nuclear engineering community has considerable potential impact to offer the world in terms of clean energy and other applications. The open source approach offers a powerful way to broadly share public knowledge to help to overcome any traditional lack of sharing.

Nuclear organizations and institutions are often under a high degree of legal and regulatory scrutiny. When it is possible to categorize a body of knowledge as open source, it becomes much more efficient from a bureaucratic point of view to establish and maintain cross-institution collaborations and discussions.

The introduction and use of OSS in the nuclear field is helping cooperation across universities, national organizations and technology vendors. OSS can be

---

[1] See https://www.unesco.org/en/open-science/inclusive-science/archiving-open-software-human-heritage

used for design, safety assessment and operation of nuclear reactors, provided that requirements related to licensing, long term maintenance and quality are met.

Many initiatives are closely related to the advent of open science and technology, including the following:

— The FAIR (findable, accessible, interoperable and reusable) principles initiative [2] establishes a set of guidelines and best practices for data management that are aimed at promoting the reuse and interoperability of data in scientific research. The initiative was first proposed in 2014 and has since been adopted by a range of organizations and initiatives in the scientific community, including the European Open Science Cloud and the National Institutes of Health. The FAIR principles state that data should be:
  - Findable: Data should be easily discoverable and accessible to both humans and machines, using metadata and persistent identifiers.
  - Accessible: Data should be available for free, or as open as possible, with clear and accessible usage licences.
  - Interoperable: Data should be structured and standardized so that they can be integrated with other data and analysed using a variety of tools and methods.
  - Reusable: Data should be well described and machine readable, with clear provenance and documentation to support reuse and reproducibility.

By promoting the adoption of the FAIR principles, the initiative seeks to improve the efficiency and impact of scientific research by making it easier to find, reuse and integrate data from diverse sources.

— The Open Source Initiative (OSI) was established in 1998 with a mission to promote and protect OSS through education, collaboration and advocacy [1, 3]. Open source code can come from various sources. It can be created and shared by individuals or it can be developed and released by organizations or communities. Often, open source projects begin as personal projects or as a solution to a specific problem encountered by developers. Additionally, some companies may release internal tools or libraries as open source to promote collaboration with the broader developer community. Finally, many open source projects are sustained by contributions from a diverse group of individuals and organizations, often spanning geographic and cultural boundaries. The concept of OSS emerged in the late 1990s, and the term 'open source' was officially associated with the OSI in 1998 [1]. However, the idea of freely sharing and collaborating on software dates to the early days of computing. Some early examples of OSS projects include the following:

- The GNU Project[2]: Launched by Richard Stallman in 1983 with the aim of creating a completely free open source operating system called GNU.
- The Linux operating system: Initially created by Linus Torvalds in 1991 while he was a student at the University of Helsinki in Finland [4]. Torvalds began developing a kernel (the core component of an operating system) as a hobby, inspired by the Unix operating system and frustrated by the licensing restrictions on proprietary software. As he worked on the kernel, Torvalds made it available for others to contribute to and improve upon, laying the groundwork for a collaborative and community driven approach to software development that remains a defining feature of the open source movement. Over time, the Linux kernel and associated software have been refined and expanded by thousands of contributors from around the world, resulting in a powerful and flexible operating system used in a wide variety of applications, from smartphones and home computers to servers and supercomputers.
- The Internet: Built on open standards and protocols that were freely available for anyone to use [5].

While these early examples were not necessarily referred to as open source at the time, they laid the foundations for the collaborative, community driven approach to software development that is now closely associated with the open source movement.

— Open design and patent initiatives demonstrate the growing trend of open design and open patents in creating more accessible, sustainable and socially responsible products and technologies. The concept of open patents is similar to that of OSS, in that it promotes collaboration and sharing to advance innovation. However, open patents are a relatively new concept and the legal mechanisms for creating and managing them are still being developed [6]. Open patents have the potential to drive innovation in a more collaborative and sustainable way by making valuable intellectual property available to a wider range of stakeholders. There are several different approaches to creating open patents, including the following:

---

² See https://www.gnu.org/

- Patent pledges: Companies or inventors pledge not to assert their patents against certain uses or users, in order to encourage innovation [7].
- Patent pools: Groups of companies agree to share their patents and allow for joint licensing, to reduce legal barriers to innovation [8].
- Defensive patenting: Companies acquire patents not for their own use, but to prevent others from using them for litigation purposes, thus making them available for defensive purposes [8].

Examples of open design initiatives include the following:

- Open Design Alliance: A non-profit consortium of companies focused on promoting open standards and interoperability in computer-aided design (CAD) software[3].
- Open Design Foundation or P2P Foundation: A non-profit organization that promotes designs for products that can be customized, repaired and upgraded, with an emphasis on environmental sustainability [9].

2.2. CHALLENGES OF TRADITIONAL SOFTWARE DEVELOPMENT IN THE NUCLEAR FIELD

Traditional methods used for software development and management in the nuclear field tend to apply the same rules to digital assets, processing software, design and patents. There is an antagonism between intellectual property and openness requirements that can be resolved only through proactive actions by the stakeholders. The analyses presented here mainly focus on processing software used as a basis to design and operate nuclear reactors. There is a distinction between the following:

(a) Elementary solutions of problems and equations (e.g. the Boltzmann transport equation, Bateman equations, Navier–Stokes equations, the Fourier equation, multiparameter interpolation);

(b) Conception of computational schemes (e.g. a consistent recipe to produce a multiparameter database starting from a cross-section evaluation, a consistent recipe to evaluate a feedback effect) for a specific class of design;

(c) A set of operating data representing an actual assembly, device or reactor.

In a proprietary code such as CASMO5 [10], items (a) and (b) are merged. In codes such as APOLLO2 [11] and DRAGON5 [12], items (a) and (b) are

---

[3] See https://www.opendesign.com

located in different components of the code system. Computational schemes are generally implemented in scripting or high level language (e.g. Python, Gibiane, CLE-2000). Openness is suitable mostly for item (a). Intellectual property restrictions occur mainly for item (c) and often lead to export control conditions in some countries. These conditions greatly complicate production use of proprietary code in consulting engineering firms, as a key production code may be forbidden for some uses.

A short term solution is to favour open source distribution of elementary solutions of problems and equations without including computational schemes related to a specific design. Digital assets, such as the Evaluated Nuclear Data File (ENDF) library [13], and coolant thermal data, such as the International Association for the Properties of Water and Steam [14] data, are already distributed without restriction. Free distribution of computer software would also facilitate engineering work. This software could be distributed as libraries of modules or classes, each dedicated to a particular operation.

2.3. OPEN SCIENCE

Adoption of open science[4] is more than developing, implementing and using OSS. A successful open science strategy involves open design and patent initiatives where computational schemes, design data and intellectual property are shared under open licences and patents. Such an outcome is only possible with the strong support of international organizations and charitable foundations. Adoption of open science could be encouraged by national and international organizations in cases where funding originates from public money.

3. OVERVIEW OF OPEN SOURCE REACTOR TOOLS AND PROJECTS

3.1. OPEN SOURCE LANDSCAPE IN NUCLEAR SCIENCE AND ENGINEERING

The main objective of this section is to summarize the open source landscape in nuclear science and engineering. For this purpose, a survey was

---

[4] See www.softwareheritage.org/

conducted involving the participants in the ONCORE Technical Meeting, as well as known developers of open source programs in the field. A summary of OSS for nuclear applications is also provided in Ref. [15]. Here, only those codes with a nuclear background or physics modelling capabilities were considered, to avoid listing large numbers of OSS (e.g. in the field of pre- and post-processing or general computational methods). Furthermore, only software that has an OSI approved licence and a publicly available repository is discussed.

In total, 40 OSS developments were identified at the ONCORE Technical Meeting (see Table 1 and Fig. 1), of which the majority are in the general nuclear engineering disciplines of thermofluid dynamics/computational fluid dynamics (CFD), structural mechanics/finite element method (FEM) and neutron transport. Some projects address disciplines in coupled form as multiphysics software (e.g. Advanced Reactor Modeling Interface (ARMI), Cardinal, Generalized Nuclear Foam (GeN-Foam), Multiphysics Object-Oriented Simulation Environment (MOOSE), TRUST/TrioCFD). Looking at CFD, approximately half the codes are derived from OpenFOAM software. There are

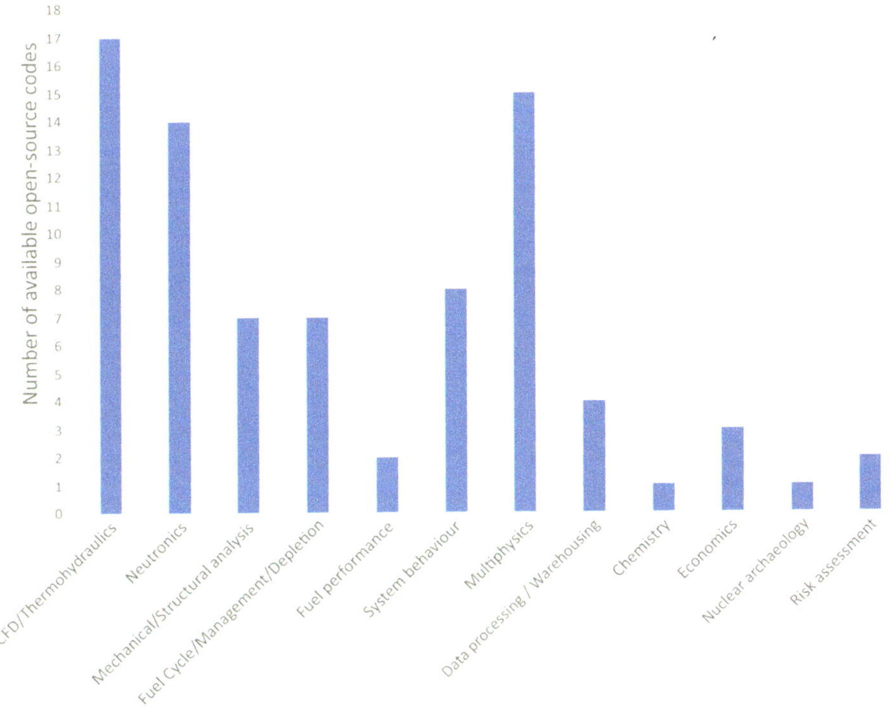

FIG. 1. Categorization of the OSS considered by the field of application (multiple answers were allowed). The categories reflect only physics and are not further grouped for specific reactor concepts (e.g. light water reactors, gas cooled reactors, molten salt reactors).

TABLE 1. SUMMARY OF THE OPEN SOURCE CODES CONSIDERED

No.	Code acronym	Licence	Developed since	Open source since	CFD/TF dynamics	Neutronics	Mechanical/str. analysis	FC management/depletion	Fuel performance	System behaviour	Multiphysics	(Nuclear) data processing	Chemistry	Economics	Nuclear archaeology	Risk assessment	Specific application
1	ARMI	Apache 2.0	2009	2019							X						
2	AURORA	LGPL v2.1	2020	2021		X					X						
3	Cardinal	GPL v2.1	2019	2021	X	X					X						
4	Code_Aster	GPL v3	1989	1999			X										
5	Code_Saturne	GPL v3	1997	1999	X												
6	containmentFOAM	GPL v3	2017	2021	X					X							LWR containment analysis
7	Cyclus	BSD 3	2010	2010				X									
8	ddtFoam	GPL v3	2009	2014	X												Combustion risk
9	DeepLynx	MIT	2019	2020						X		X					Data storage
10	DAGMC	BSD 3	2003	2003		X											
11	DRAGON5	LGPL v3	2014	2016		X											
12	ENRICO	MIT	2019	2019							X						
13	flameFoam	GPL v3	2021	2021	X												Combustion risk

TABLE 1. SUMMARY OF THE OPEN SOURCE CODES CONSIDERED (cont.)

No.	Code acronym	Licence	Developed since	Open source since	CFD/TF dynamics	Neutronics	Mechanical/str. analysis	FC management/depletion	Fuel performance	System behaviour	Multiphysics	(Nuclear) data processing	Chemistry	Economics	Nuclear archaeology	Risk assessment	Specific application
14	FLEXPART	GPL v3	1998	2013	X												Atmospheric dispersion
15	GeN-Foam	GPL v3	2014	2016	X	X	X				X						
16	KOMODO	MIT	2017	2020		X					X						
17	Modelica	BSD 3	1996	2000						X							
18	MOOSE	LGPL 2.1	2008	2014	X		X			X	X		X				
19	NEK5000	Individual	1987	2008	X												
20	nekRS	BSD 3	2019	2019	X												
21	NEORL	MIT	2020	2020		X		X									
22	NJOY	BSD 3	2013	2016								X					
23	NCET	MIT	2021	2023										X			
24	OFFBEAT	GPL v3	2017	2022				X	X								
25	ONIX	MIT	2019	2019				X							X		

TABLE 1. SUMMARY OF THE OPEN SOURCE CODES CONSIDERED (cont.)

No.	Code acronym	Licence	Developed since	Open source since	CFD/TF dynamics	Neutronics	Mechanical/str. analysis	FC management/depletion	Fuel performance	System behaviour	Multiphysics	(Nuclear) data processing	Chemistry	Economics	Nuclear archaeology	Risk assessment	Specific application
26	OpenFOAM (Foundation)	GPL v3	2004	2011	X												
27	OpenFOAM (ESI)	GPL v3	2004	2011	X												
28	OpenMC	MIT	2011	2012		X		X			X						
29	OpenMOC	MIT	2013	2013		X											
30	OpenModelica	GPL v3	1997	2008						X							
31	OpenPRA (LOGOS, CUDD, EMRALD, SCRAM, XFTA)	MIT, GPL	2018							X						X	
32	PyNE	BSD 2	2010	2010								X					
33	RAVEN	Apache v2.0	2017	2017						X	X			X		X	Optimization, PRA, UQ and SA
34	SALOME	GPL v3	1998	1998							X						
35	SCIANTIX	MIT	2016	2018			X	X									Fission gas release
36	SCONE	MIT	2018	2020	X												

TABLE 1. SUMMARY OF THE OPEN SOURCE CODES CONSIDERED (cont.)

No.	Code acronym	Licence	Developed since	Open source since	CFD/TF dynamics	Neutronics	Mechanical/str. analysis	FC management/depletion	Fuel performance	System behaviour	Multiphysics	(Nuclear) data processing	Chemistry	Economics	Nuclear archaeology	Risk assessment	Specific application
37	TALYS	GPL v3 / CECILL A	2004	2004								X					
38	TFEL/MFront	GPL v3	2006	2014			X				X						
39	TrioCFD	BSD 3	1995	2015	X		X				X						
40	TRUST	BSD 3	1995	2015	X	X	X				X						
Open source licence but not publicly available at the time of writing																	
41	CORE SIM	GPL v3	2011			X											
42	explosionDynamicsFoam	GPL v3	2019	request	X												Combustion risk
43	HCP	MIT	2010	request	X	X		X		X	X						HTGR analysis
44	OpenFOAM_RCS	GPL v3	2020	request	X												LWR cooling
45	VANGARD	Dual GPL v3 + commercial	2020	request		X		X			X						

Note: BSD — Berkeley software distribution; CFD — computational fluid dynamics; DAGMC — Direct Accelerated Geometry Monte Carlo; FC — fuel cycle; GPL — GNU General Public Licencs; HCP — high temperature reactor code package; HTGR — high temperature gas cooled reactor; LGPL — GNU Lesser General Public Licencs; LWR — light water reactor; NCET — Nuclear Cost Estimation Tool; NEORL — NeuroEvolution Optimization with Reinforcement Learning; OFFBEAT — OpenFOAM Fuel Behaviour Analysis Tool; PRA — probabilistic risk assessment; SA — systems analysis; str. analysis — structural analysis; TF — thermofluid; UQ — uncertainty quantification.

at least a few developments in almost all other fields related to nuclear science and engineering (e.g. system behaviour, fuel management, structural analysis, risk assessment, economics). For specific subjects such as radiochemistry, economics, risk assessment and fuel performance, only very few software developments were reported.

In the codes mentioned in Table 1, there is no unique definition of open source 'frameworks', 'projects', 'activities' or 'codes', and thus no further categorization is attempted. However, looking at the identified projects, there are several computer codes with a quite specific purpose, as well as a few frameworks (e.g. OpenPRA, ARMI, MOOSE) that couple or integrate different codes to achieve a broader application. Further quantitative criteria describing the efforts in a sustainable development or the number of developers may be used in the future to distinguish between the different classes of open source developments.

3.2. DEVELOPMENT AND LICENCES

Approximately 50% of the codes in Table 1 were first developed within the past ten years, while the earliest developments date back to the 1990s (see Fig. 2, left). They were made open source directly from the beginning or after a

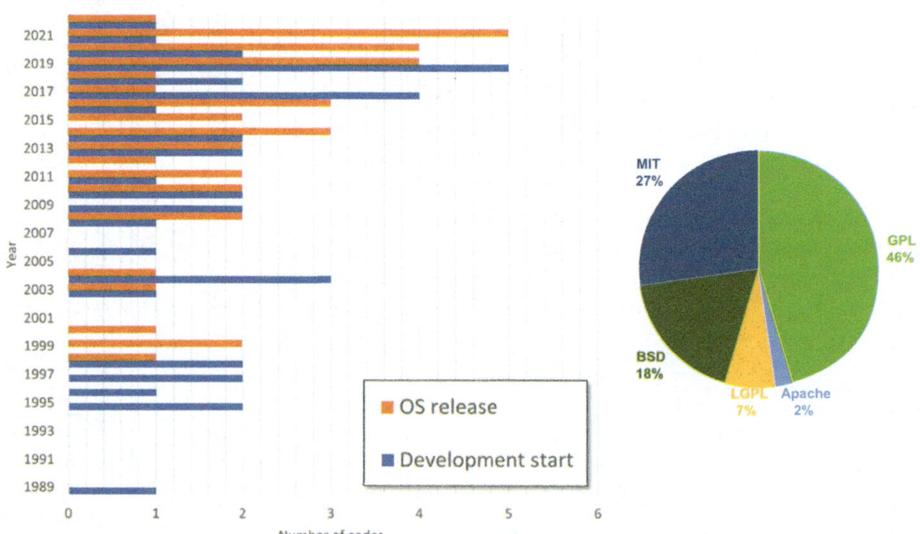

FIG. 2. Development and release history (left), chosen open source licence (right). BSD — Berkeley source distribution 3-clause licence; GPL — GNU General Public License; LGPL — GNU Lesser General Public License; MIT — Massachusetts Institute of Technology licence; OS — open source software.

consolidation phase. The OSI licence types that are primarily used are the copyleft GNU General Public License (GPL) version 3, or the GNU Lesser General Public License (LGPL), or the more permissive Massachusetts Institute of Technology (MIT) licence, the Berkeley source distribution (BSD) 3-clause licence or the Apache licence (see Fig. 2, right). While many developments originate from universities or other research organizations, the SALOME platform, Code_Aster and Code_Saturne from Électricité de France (EDF) and ARMI from TerraPower are examples of industry driven developments.

3.3. MAINTENANCE

Most of the codes considered here are actively maintained; for some, only bug fixes are provided and a few codes are outdated. The frequency of updates varies from weekly commits to sporadic synchronization with the corresponding development lines. While agile practices enable more direct feedback between users and developers, time constraints of developers may hinder frequent interaction with the user group. The efforts for maintenance and development vary considerably, ranging from a single developer to a full team (of no more than 20 developers) and/or a community. The developer teams cover the full range of qualifications, from students to scientists or software engineers. The funding of code maintenance is often reported to be gathered from multiple R&D projects that focus on the application of the software. Maintenance itself is often reported as 'non-scientific' work and thus not covered by direct funding of the laboratories or universities.

Almost all software is developed and distributed via a code version management platform (e.g. GitLab[5], GitHub[6], SourceForge[7], Bitbucket[8]), while dedicated web sites are also maintained in some cases. To aid maintenance and enable QA, most developments rely on the features provided along with

---

[5] GitLab is a web based DevOps life cycle tool that provides a Git repository manager with a wiki, issue tracking and continuous integration and deployment pipeline features using an open source licence.

[6] GitHub is a web based platform used for version control and collaboration on software development projects. It offers distributed version control and source code management for the functionality of Git, as well as its own features and user interface.

[7] SourceForge is a web based platform that offers services to software developers. It provides a centralized location for hosting and managing OSS projects.

[8] Bitbucket is a web based platform for hosting Git and Mercurial repositories. It provides tools for code collaboration, version control and project management, similar to GitHub and GitLab.

the version control system (i.e. ticket based bug/issue reporting, continuous integration along with manual or automatic testing (pipelines)).

For larger code developments/teams, there are several reports on the application of: (a) formalized review procedures; (b) code style guides and comprehensive automatic testing (i.e. unit tests, such as pFUnit[9], up to regression testing with more than 1000 verification and validation (V&V) cases); as well as (c) quality control (e.g. Valgrind[10]). Most tools are developed for a specific operating system, but some packages are maintained as native versions or docker containers for the most prominent versions/distributions of Linux, Windows and macOS or OS X.

Feedback and interaction with users are mostly via issue/bug trackers but occasionally via dedicated email, forums or chat software. The documentation of the codes is primarily done through commented source codes, automatic generation of documentation (using e.g. Doxygen[11]), the development history (e.g. version control system commit messages), wiki pages or and specific web services (e.g. readthedocs[12], GitHub and GitLab pages). An overview of the maintenance approaches is given in Section 5, which explores best practices.

3.4. DESCRIPTIONS OF THE APPLICATION, FUNCTIONALITY AND SCOPE OF THE CODES

The following subsections provide a summary of the available OSS packages and frameworks.

3.4.1. Advanced Reactor Modelling Interface

ARMI[13] [16] provides a high level nuclear reactor data model and a variety of nuclear specific utilities and analyses designed to streamline typical reactor design and analysis tasks. Given a description of a reactor and material

---

[9] pFUnit is an open source testing framework specifically designed for the Fortran programming language. It helps Fortran developers write and execute unit tests to ensure the correctness and reliability of their code.

[10] Valgrind is an open source programming tool suite primarily used for memory debugging, memory leak detection and profiling. It is widely used in the software development industry, especially for C and C++ programs, but it also supports other languages.

[11] Doxygen is an open source documentation generator tool primarily used for documenting source code. It supports multiple programming languages, including C, C++, Java and Python.

[12] readthedocs is a popular documentation hosting platform that automates the process of building, versioning and hosting documentation for software projects.

[13] Repository: https://github.com/terrapower/armi

properties library, it can compute as-built dimensions and composition, provide homogenized number densities, plot depletion matrices, step through time, perform fuel management and so on, as informed by a user specified (and user or community provided) set of physics plug-ins. The framework itself does not include physics modules beyond homogenizations and thermal expansions but it is expected that plug-ins will be used from different subfields. For example, given the composition and dimensions from an ARMI model, a neutronics plug-in could write neutronics input, run a flux and power solver and add flux data back into the reactor model. A depletion plug-in could then pick this up from the central model and compute reaction rates. Temperature distributions could then be computed. The central data model provides a common language that all these plug-ins and their associated physics models can work with. Users often take a bespoke physics tool and write wrapper code to translate its inputs and outputs into the ARMI data format. Therefore, any code can be made to couple and automate alongside other codes, providing rapid full scope design analysis. The ARMI framework has historically been funded primarily by sodium cooled fast reactor projects. As such, its capabilities are fine-tuned for that specific application. The nuclear reactor design company TerraPower has modelled molten salt reactors with it, and other companies are working on high temperature gas cooled reactors. Additional work is needed before effective models of light water reactors (LWRs) and other reactors are fully supported.
Methodology: Data model, coupling system and set of automation utilities.

3.4.2. AURORA

AURORA[14] [17] (A Unified Resource for OpenMC (fusion) Reactor Applications) combines the Monte Carlo neutron transport calculations from OpenMC with the finite element analysis calculations supported by the MOOSE framework, intended for the modelling of tokamak physics. Now MOOSE supports the Heat Conduction and Tensor Mechanics modules, intended to model the increase in temperature arising from heat deposited from neutrons, and subsequent thermal expansion and density changes, respectively. Support for additional modules is expected. Coupling Monte Carlo neutron transport code via OpenMC to MOOSE finite element analysis in-built modules targets heat conduction and thermal expansion.
Methodology: Coupling code between Monte Carlo code OpenMC and the MOOSE framework.

---

[14] Repository: https://github.com/aurora-multiphysics/aurora

3.4.3. Cardinal

Cardinal[15] [18, 19] is an application in the MOOSE framework that couples OpenMC and NekRS. Its coupling scheme allows it to transfer information from a neutronics simulation to thermal hydraulics simulations in a variety of ways, leveraging both a cell averaged and an unstructured mesh approach on the neutronics side. It can also couple to other codes in the MOOSE framework, such as SAM and BISON, for more extensive analysis.
Methodology: MOOSE application that wraps non-MOOSE codes (i.e. OpenMC and NekRS) into the framework.

3.4.4. Code_Aster

Code_Aster[16] is a finite element analysis solid mechanics OSS developed by EDF to tackle challenges related to power generation. It is designed specifically for structural mechanics in the field of nuclear engineering. The software has been developed and kept updated for more than 30 years and is used throughout the world by numerous engineers who are part of a large open source community. The tool is widely used for the evaluation and maintenance of power plants and electrical networks and covers a large range of current applications: mechanical and thermal analysis; 3-D linear or non-linear, static or dynamic analysis; pressure vessels; civil engineering; porous media; fracture mechanics and fatigue; soil mechanics; welding; and additive manufacturing.
Methodology: Finite element framework.

3.4.5. Code_saturne

Code_saturne[17] is a free OSS developed and released by EDF to perform CFD applications. It solves the Navier–Stokes equations for 2-D, 2-D axisymmetric and 3-D flows, steady or unsteady, laminar or turbulent, incompressible or weakly dilatable, isothermal or not, with scalar transport if required. Several turbulence models are available, from Reynolds averaged models to large eddy simulation models. In addition, a number of specific physical models are also available as modules: semitransparent radiative transfer, particle tracking with Lagrangian modelling, electric arcs, weakly compressible flows, atmospheric flows and rotor–stator interaction for hydraulic machines. For the nuclear industry it is widely used for 3-D steam line break studies; pressure

---

[15] Repository: https://github.com/neams-th-coe/cardinal

[16] Repository: http://www.code-aster.org

[17] Repository: https://www.code-saturne.org

and temperature assessment in the containment; modelling of pressurized thermal shock; modelling of loss of coolant accident; and core design. Code_saturne serves as a basis for Code Neptune (not an OSS) for the modelling of two phase flow for applications such as the assessment of critical heat flux and of severe accidents.

Methodology: Finite element based CFD.

3.4.6. containmentFOAM

containmentFOAM[18] [20–22] is a multispecies and multiphysics toolbox based on OpenFOAM-9. It is designed to efficiently simulate transport processes within confined domains, such as a nuclear reactor containment during a severe accident. It may be used in technical scale safety assessments. It contains submodels to account for pressurization, heat and mass transfer in buoyant flows, gas radiation heat transport, combustible gas (H_2/CO) mixing and mitigation, as well as aerosol particle transport. Currently models are tailored for large dry pressurized water cooled reactor (PWR) containments in Germany. Extension is needed for other containment types utilizing different safety measures (e.g. spray systems) or different design conditions (e.g. boiling water reactor or small modular reactor containments).

Methodology: Finite volume method (FVM) based CFD approach to resolve gas transport and mixing. Monte Carlo emission reciprocity method for thermal radiation transport. Technical and safety systems are modelled either using homogenized (porous media) models or by code coupling to external codes (e.g. simulators of passive autocatalytic recombiners. Thermal radiation transport is modelled by a Monte Carlo neutron transport code built upon OpenFOAM libraries.

3.4.7. Cyclus

Cyclus[19] [23] provides an agent based framework that enables discrete facilities to interact and exchange commodities (typically nuclear material). These facilities can be represented by arbitrarily complex plug-in modules (e.g. Cycamore).

Methodology: The framework itself provides a market-like mechanism, known as the dynamic resource exchange, which (a) collects requests for materials from consumers, (b) collects offers to provide those materials from producers, (c) collects consumer preferences for each possible offer, and

---

[18] Repository: https://go.fzj.de/containmentFOAM
[19] Repository: https://github.com/cyclus/cyclus

(d) solves a network flow problem according to these preferences. The models for both physical behaviour and dynamic resource exchange interaction behaviour are encapsulated in plug-in modules for specific facilities, such as modules included in the Cycamore library of minimum working facility models.

3.4.8. Direct Accelerated Geometry Monte Carlo

Direct Accelerated Geometry Monte Carlo (DAGMC)[20] [24, 25] is a C++ software library used to provide accelerated geometry operations on a high fidelity tessellated representation of a CAD geometry that can be integrated with most Monte Carlo radiation transport codes. Integrations with OpenMC, MCNP6, FLUKA and Geant4 are actively maintained. Proof of principle integrations with Shift and Tripoli also exist.
Methodology: Bounding volume hierarchy accelerations of ray tracing operations on facets.

3.4.9. ddtFOAM

The ddtFOAM model[21] [26] is built on the OpenFOAM framework; it focuses on deflagration to detonation transition and is limited to fast hydrogen flames. The computational model is based on the formulation of a reaction progress variable and accounts for both deflagrative flame propagation and autoignition. The model can simulate the deflagration to detonation transition without resolving all microscopic details of the flow. It works on relatively coarse grids and shows good agreement with experiments.
Methodology: FVM.

3.4.10. DeepLynx

DeepLynx[22] [27–29] is a general purpose data warehouse where data are stored in a graph-like format following a user defined ontology. Through GraphQL, users and applications can request exactly the data that they need by using client-side defined queries. Engineering projects often operate using siloed tools and disparate teams where connections across design, procurement and construction systems are translated manually or over brittle point-to-point integrations. The manual nature of data exchange increases the risk of undetected ('silent') errors in the project design, with each silent error potentially propagating

---

throughout the system. Such cascading errors can create uncontrolled risks during construction, leading to significant delays and cost overruns. DeepLynx provides an integrated platform across all phases of design and operation, mitigating these risks in mega projects.

Methodology: DeepLynx uses a PostgreSQL database to store data as a graph.

3.4.11. DRAGON5

DRAGON5[23] [30, 31] has been developed for lattice (fundamental mode) and reflector (non-fundamental mode) calculations and models. It has a potential application to Canada deuterium–uranium (CANDU) reactors and PWRs. DRAGON5 is a collection of independent modules. Production calculation schemes for CANDU reactors and PWRs are not included in the LGPL distribution.

Methodology: Solution of the Boltzmann equation (e.g. method of characteristics, collision probabilities, interface current method, discrete ordinates method). Solution of Bateman equations. Resonant self-shielding model based on the subgroup method.

3.4.12. Exascale Nuclear Reactor Investigative Code

The Exascale Nuclear Reactor Investigative Code (ENRICO)[24] [32] is a multiphysics framework designed for analysis of LWRs. It provides a framework for coupling neutronics and thermal hydraulics simulations that is code agnostic, with specific support for the OpenMC, SHIFT and NekRS codes as well as an internal subchannel solver. An experimental branch also contains support for coupling with OpenFOAM.

Methodology: Monte Carlo, thermal hydraulics, CFD, FEM, FVM.

3.4.13. flameFOAM

flameFOAM[25] [33–35] is an OpenFOAM based combustion solver. It implements a progress variable approach and a turbulent flame closure model for premixed combustion simulation. Turbulent flame speed is defined as a function of laminar flame speed and turbulence parameters and is evaluated using one of three different correlations. Laminar flame speed can be set as a

---

[23] Repository: http://merlin.polymtl.ca

[24] Repository: https://github.com/enrico-dev/enrico

[25] Repository: https://github.com/flameFoam/flameFoam

constant or calculated using a laminar flame speed correlation. The standard compressible OpenFOAM solvers rhoPimpleFoam, buoyantPimpleFoam and chtMultiRegionFoam are the basis for the solver. flameFOAM is currently applicable to premixed turbulent combustion of homogeneous hydrogen–air mixtures only. Simulations employ simplified approaches, which allow one to run models with affordable computational cost and at the same time provide sufficiently accurate results. The employed approaches are Reynolds averaged Navier–Stokes modelling of turbulence and combustion models based on progress variable, turbulent flame closure and turbulent flame speed correlations. An optional artificial neural network model for laminar burning velocity is also included.

Methodology: Entirely based on the FVM on unstructured meshes, with parallelization based on domain decomposition.

3.4.14. FLEXPART

FLEXPART[26] [36–39] is a Lagrangian dispersion model created to compute the extended range and mesoscale dispersion of air pollutants from various point sources, such as after an accident at a nuclear power plant, in both forward and backward analysis. FLEXPART has evolved into a comprehensive tool for atmospheric transport modelling and analysis. Its application fields have been extended from air pollution studies to other topics where atmospheric transport plays a significant role (e.g. exchange between the stratosphere and troposphere, the global water cycle). It has grown into a genuine community model, currently used by at least 35 groups across 14 different countries, with applications in both operational and research contexts.

Methodology: Lagrangian modelling of particles in atmospheric field.

3.4.15. GeN-Foam

GeN-Foam[27] [40–45] is a multiphysics solver for reactor analysis based on OpenFOAM. It can solve (coupled or selectively) for the following:

— Neutronics, with models for point kinetics, diffusion (transient and eigenvalue), adjoint diffusion (only eigenvalue), SP_3 (transient and eigenvalue) and discrete ordinates (only eigenvalue);

— One-phase thermal hydraulics, according to both Reynolds averaged Navier–Stokes CFD and porous-medium, coarse-mesh approaches (the two approaches can be combined in the same mesh);
— Two-phase porous-medium thermal hydraulics, according to an Euler–Euler model, with full capabilities for sodium cooled fast reactors, pre-critical heat flux capabilities for LWRs, and preliminary and post-critical heat flux capabilities;
— Thermomechanical analysis based on linear thermoelasticity, which can be used to evaluate deformations in a reactor core. Such deformations are used to modify the meshes for thermal hydraulics and neutronics.

GeN-Foam can be coupled with the Serpent Monte Carlo code using the Serpent multiphysics interface as well as functional mock-up objects via the FMU4FOAM library. In addition, it can be coupled with any existing OpenFOAM solver either at the level of the source code, via the standard OpenFOAM application programming interface (API), or by making use of the runtime multiapplication coupling available in OpenFOAM.
Methodology: Entirely based on the FVM on unstructured meshes, with parallelization based on domain decomposition.

3.4.16. KOMODO

KOMODO[28] [46, 47] is an open source nuclear reactor simulator that solves both static and transient neutron diffusion equations for 1-D, 2-D or 3-D reactor problems in Cartesian geometry. Currently KOMODO uses a semianalytic nodal method by default for the spatial discretization of the neutron diffusion equation. The theta method is used for the time discretization.

Komodo includes the following features:

— It solves both static and transient core problems with the possibility to couple with thermal hydraulics.
— It performs forward, adjoint and fixed source simulations.
— It performs calculations using branched cross-section data.
— It executes searches for critical boron concentration.
— It simulates rod ejection or reactivity initiated accidents.
— It uses the multigroup approach to solve neutron diffusion equations.

---

[28] Repository: https://github.com/imronuke/KOMODO

— It solves problems with assembly discontinuity factors.

— It uses two-node problem non-linear iteration to accelerate a coarse mesh finite difference solver.

Methodology: The initial version of KOMODO employed a nodal expansion method based on the response matrix formulations. The transverse integrated leakages were approximated by the quadratic transverse leakage approximation. The transient and thermal modules were also added to enable KOMODO to solve time dependent problems with thermal hydraulics feedback. The transient diffusion equation was solved with a fully implicit method, and thermal hydraulics solutions were obtained by solving mass and energy conservation equations in an enclosed channel.

In early 2020, coarse mesh finite difference acceleration with two-node problems and non-linear iteration procedures was implemented in KOMODO to accelerate the calculations. New nodal kernels — a polynomial nodal method and a semianalytic nodal method — were also added to KOMODO. Users also have the option to perform exponential flux transformation and to set a theta value for time dependent problems. It is also possible to perform calculations with branched cross-sections or parametric macroscopic cross-sections (i.e. sets of cross-sections for various thermal hydraulics parameters, boron concentration and control rod position).

3.4.17. Modelica/OpenModelica

Modelica[29] [48] is a modern modelling language used for the simulation of physical and engineering systems. It features different characteristics that also make it suitable for use as a system code in the nuclear field, namely:

— It is object oriented, granting a hierarchical structure, inheritance, abstraction and encapsulation that provide modularity, openness and efficiency to the modelling tool.

— It is declarative, focusing on what the models describe rather than how the model is solved.

— It is equation based, solving differential algebraic equation systems rather than ordinary differential equations for a better implementation of engineering first principles.

— It features acausal modelling, not specifying the classic input–output relationship but granting a more flexible and efficient data flow.

---

[29] Repository: https://github.com/modelica, https://openmodelica.org

— It is component oriented, allowing a description of single system components (or objects) in terms of physical equations and handling the connections among them through standardized interfaces (or connectors).

The last of these points — in addition to providing a more realistic representation of the plant — allows the creation of libraries of power plant components that can be reused not only in the same model but in other power plant configurations. Several validated libraries are already available that the nuclear community can take advantage of[30].

Other advantages of the use of Modelica are the following:

— The presence of already validated components, especially the conventional non-nuclear-specific ones.
— The possibility to develop ad hoc components as an extension of some base classes. In addition, ad hoc correlations (e.g. for pressure drops and heat transfer coefficients) suitable for nuclear plant conditions can be easily adopted and implemented, especially for the different flow regimes. In the framework of the open data paradigm — especially in terms of the separate effect of integral experiments — this can constitute a remarkable step for the thermal hydraulics community.
— OSS allows the investigation of the modelling approach used for each component, possibly improving or tuning the model approach to the specific use (e.g. avoiding the risk of over-modelling). OSS will also help in the V&V process, as errors will be easier to detect.
— The possibility to couple the Modelica based model with other modelling approaches through the Functional Mock-up Interface.

Modelica is currently used in different fields, such as the automotive industry, robotics, thermal hydraulics and mechatronics. Several applications also involve the nuclear simulation field as a power plant simulator for control purposes for the ALFRED reactor, the IRIS reactor, the molten salt fast neutron reactor and the P4 PWR. Modelica shares the same limitation as the system code approach (mainly 1-D). Modelica is a modelling language and requires a simulation environment to be used. This is the main scope of OpenModelica, an open source and non-proprietary modelling and simulation environment for Modelica. OpenModelica implements front end and back end (i.e. pre-optimization, causalization and post-optimization) compiling steps, as well as code generation and executable simulation. In addition, its scope also includes

---

[30] Validated libraries are available here: https://modelica.org/libraries.html

model based analysis such as optimization, debugging and co-simulation of Functional Mock-up Interface based models.

Methodology: Solving differential algebraic equations.

3.4.18. Multiphysics Object Oriented Simulation Environment

MOOSE is a fundamental framework for building new multiphysics simulation tools and coupling existing ones. In addition, the MOOSE framework[31] [49, 50] contains a set of physics modules that cover a range of simulations, from heat conduction, solid mechanics and chemistry to massively parallel raytracing.

Methodology: Although MOOSE was originally an FEM code, it now supports many formulations, including the FVM. In addition, there are over 30 pluggable interfaces that allow the development of custom numerical methods. Many different nuclear codes have been built using it, including codes for fuel performance, material degradation, neutronics, core fluid flow, systems analysis and corrosion analysis.

3.4.19. NEK5000, nekRS

High order methods have the potential to overcome the current limitations of standard CFD solvers. For this reason, the spectral element code NEK5000[32] [51] has been continuously developed for over 35 years. It features state of the art, scalable algorithms that are fast and efficient on platforms ranging from laptops to the world's fastest computers. Applications span a wide range of fields, including fluid flow, thermal convection, combustion and magnetohydrodynamics.

nekRS is a next-generation open source Navier–Stokes solver based on the spectral element method targeting classical processors and accelerators such as graphic processing units (GPUs). The code started as a fork of libParanumal in 2019. For API portable programming, open concurrent computing abstraction is used.

The capabilities of NEK5000 and nekRS include the following:

— Incompressible and low Mach number Navier–Stokes and scalar transport;
— Continuous Galerkin spectral element method using curvilinear conformal hexahedral elements;
— Variable time step second and third order semi-implicit time integration;

---

31 Repository: https://github.com/idaholab/moose
32 Repository: https://nek5000.mcs.anl.gov

— MPI+X hybrid parallelism supporting a central processing unit (CPU), compute unified device architecture, heterogenous compute interface for portability and open computing language;
— Various boundary conditions;
— Conjugate fluid–solid heat transfer;
— Large eddy simulation and Reynolds averaged Navier–Stokes turbulence models;
— Arbitrary Lagrangian–Eulerian moving mesh;
— VisIt and ParaView support for data analysis and visualization.

3.4.20. NJOY

The NJOY Nuclear Data Processing System[33], version 2016 [52], is a comprehensive computer code package for producing pointwise and multigroup cross-sections and related quantities from evaluated nuclear data in the ENDF4 through ENDF6 legacy card image formats. NJOY works with evaluated files for incident neutrons, photons and charged particles, producing libraries for a wide variety of particle transport and reactor analysis codes.

3.4.21. Nuclear Cost Estimation Tool

NCET[34] [53–55] was developed for the capital cost and schedule estimation of water cooled and high temperature gas reactors. Its functionalities include modularization, advanced construction methods and learning rates, as well as built-in sensitivity and uncertainty analysis techniques. The tool will provide a detailed bottom-up evaluation of emerging nuclear technologies.
Methodology: Excel based input with Python wrapper and Monte Carlo sampler for uncertainty and sensitivity analysis.

3.4.22. NeuroEvolution Optimization with Reinforcement Learning

NEORL[35] [54–56] offers a global optimization interface of state of the art algorithms in the field of evolutionary computation, neural networks through reinforcement learning and hybrid neuroevolutionary algorithms. NEORL features a diverse set of algorithms, a user friendly interface, parallel computing support, automatic hyperparameter tuning, detailed documentation and demonstration of applications in mathematical and real world engineering

---

optimization. NEORL encompasses various optimization problems, from combinatorial, continuous, mixed discrete–continuous optimization to high dimensional, expensive and constrained engineering optimization. NEORL is being tested in a variety of engineering applications relevant to low carbon energy research in solutions to climate change. Examples include nuclear reactor control, nuclear fuel optimization, mechanical and structural design optimization and fuel cell power production. NEORL's limitations include its focus on single objective optimization, where special techniques are needed to convert multiobjective optimization problems to the equivalent single objective problems. Ongoing efforts focus on adding algorithms for multiobjective optimization.

3.4.23. OpenFOAM Fuel BEhavior Analysis Tool

OFFBEAT[36] [57–61] is a 3-D finite volume nuclear fuel performance code based on the OpenFOAM C++ library. It was developed according to a cell cantered finite volume framework. This is combined with a framework for thermal analysis and with numerical developments concerning the treatment of the gap heat transfer and contact, based on a mapping algorithm that allows the use of independent non-conformal meshes for fuel and cladding. The code considers the temperature and burnup dependence of the material properties, and it can model fuel densification, relocation, swelling, growth, fission gas release, creep, plasticity and other relevant fuel behaviour phenomena.

OFFBEAT was created as a joint development by the Laboratory of Reactor Safety at the Swiss Federal Institute of Technology Lausanne and the Laboratory for Reactor Physics and Thermal-Hydraulics at the Paul Scherrer Institute. It is currently under development in the framework of the Euratom's OperaHPC project.

OFFBEAT is equipped with most of the features expected from a fuel performance code, such as the following:

— Experimental correlations for the temperature dependent and burnup dependent material properties;
— Semi-empirical models for fuel relocation, cracking, densification and irradiation growth;
— Models for rate independent plasticity and creep;
— The SCIANTIX code as a state of the art, mechanistic model for fission gas release and fission gas swelling;

---

[36] Repository: https://gitlab.com/foam-for-nuclear/OFFBEAT

— A simplified neutronics module derived from the TRANSURANUS burn-up model to calculate radial power profiles and isotopic evolution;
— A gap conductance model derived from the FRAPCON code but extended to a multidimensional framework;
— Appropriate methodologies to simulate gap closure and the fuel pellet thermomechanical contact;
— Scripts and tools that simplify the creation of geometries and post-processing of data.

Methodology: OFFBEAT is based on the C++ library OpenFOAM, which employs the FVM to discretize and solve systems of partial differential equations.

3.4.24. ONIX

ONIX[37] (OpeN IsotopiX) [62] is a validated nuclear depletion software. ONIX models the transmutation of irradiated isotopes to a neutron flux and the decay of radioactive isotopes. The software can be coupled with the open source Monte Carlo transport code OpenMC and offers a full reactor physics package to simulate nuclear reactors. The code is open source and written in Python.
Methodology: To solve the depletion equation, ONIX uses a 16th order Chebyshev rational approximation method. The code can be used with any nuclear data library as long as the libraries are converted into a local format.

3.4.25. OpenFOAM

OpenFOAM[38] [63, 64] has an extensive range of features to simulate a range of situations — from turbulent flows in automotive aerodynamics to fires and fire suppression in buildings — involving combustion, chemical reactions, heat transfer, liquid sprays and films. It includes tools for meshing in and around complex geometries, for data processing and visualization. Almost all computations can be executed in parallel to take full advantage of today's multicore processors and multiprocessor computers.

There are two major versions, one maintained by the OpenFOAM Foundation[39] and the other maintained by ESI[40]. The two versions are increasingly

---

[37] Repository: https://github.com/jlanversin/ONIX

[38] Repository: https://github.com/OpenFOAM/OpenFOAM-dev (OpenFOAM Foundation); https://develop.openfoam.com/Development/openfoam (ESI)

[39] Available at: www.openfoam.org

[40] Available at: www.openfoam.com (https://www.esi-group.com/)

deviating from each other, which makes it difficult to transfer models and methods from one to the other. While the OpenFOAM Foundation focuses on a thorough redesign and refactoring of the source code and integration of advanced multiphase flow modelling capabilities, the ESI version has a broader range of specific models and usability improvements.
Methodology: FVM.

3.4.26. OpenMC

OpenMC[41] [65–67] is a Monte Carlo simulation code designed by the community for simulating the transport of neutrons and photons. Its capabilities include executing fixed source, k eigenvalue and subcritical multiplication calculations on models created using constructive solid geometry or CAD representation. OpenMC is equipped to handle both continuous energy and multigroup transport. The continuous energy particle interaction data are stored in a native HDF5 format, which can be derived from ACE files generated by NJOY. The software allows for parallelism through a hybrid Message Passing Interface (MPI) and OpenMP programming model.

OpenMC includes (a) a Monte Carlo particle transport solver, (b) a depletion solver and (c) a rich Python API that includes nuclear data functionality.
Methodology: Monte Carlo transport, Chebyshev rational approximation method for depletion.

3.4.27. OpenMOC

OpenMOC[42] [68] is a method of characteristics neutral particle transport code for reactor physics criticality calculations. It is capable of simulating 2-D assembly and full-core reactor models based on constructive solid geometry with second-order surfaces, as well as 3-D extruded geometries. High performance parallel solvers for multicore CPUs and GPUs (2-D only) are also provided, as well as domain decomposition. The solver includes both the flat source and the linear source method of characteristics implementations, and acceleration is provided using coarse mesh finite difference acceleration.
Methodology: Method of characteristics, flat and linear sources, coarse mesh finite difference acceleration.

---

[41] Repository: https://github.com/openmc-dev/openmc
[42] Repository: https://github.com/mit-crpg/OpenMOC

3.4.28. OpenPRA

The OpenPRA initiative[43] intends to provide a more informative view of the actual systems, components and interactions that probabilistic risk assessment (PRA) models represent to help decision makers. The following codes are provided in the framework:

— SCRAM: probabilistic risk analysis tool (e.g. fault tree analysis, event tree analysis).[44]
— EMRALD (event modelling risk assessment using linked diagrams): used for researching the capabilities of dynamic probabilistic risk assessment.[45]
— CUDD (Colorado University decision diagram): developed for the manipulation of decision diagrams. It supports binary decision diagrams, algebraic decision diagrams and zero suppressed binary decision diagrams.[46]
— XFTA: a powerful and efficient calculation engine for fault trees and related models.[47]
— The codes are interlinked in a web application.[48]

Methodology: Event tree, fault tree, probabilistic risk assessment, binary.

3.4.29. PyNE

The tool PyNE[49] [69, 70] can translate input/output data from one nuclear engineering tool to another, often through a canonical internal representation. PyNE can also include reference implementations of workflows to support complex analysis.
Methodology: The majority of PyNE consists of methods to read output and construct input for different nuclear engineering analysis tools. Frequently, PyNE has a canonical internal representation that is compatible with the set of supported formats to allow one to many implementations to be used in a many to many mode. Although it began mostly in Python, much of PyNE is now implemented in C++ with Python bindings to allow the identical implementation to be used in compiled codes as the Python tools.

---

[43] https://openpra.org. Repository: https://gitlab.openpra.org
[44] Available at: https://gitlab.openpra.org/scram/scram
[45] Available at: https://gitlab.openpra.org/publics/EMRALD
[46] Available at: https://gitlab.openpra.org/publics/cudd
[47] Available at: https://gitlab.openpra.org/publics/xfta
[48] Available at: https://app.openpra.org
[49] Repository: https://github.com/pyne/pyne

3.4.30. RAVEN

RAVEN[50] (Risk Analysis Virtual ENvironment) [71–73] is an open source flexible computational framework designed to perform statistical analysis workflows that explore the response of simulation models (e.g. for uncertainty propagation, model optimization, model calibration and model validation). RAVEN offers a 'bring your own model' approach to perform these analyses or, alternatively, surrogate models that can be created using RAVEN. With RAVEN, users can explore the response of numerical models to a variety of initial and operating conditions, analysing the resulting data using machine learning, data mining and artificial intelligence algorithms. These analysis workflows can be executed on multiple operating systems and hardware configurations, ranging from laptops to high performance computing (HPC) environments.

RAVEN also provides a plug-in interface that has already been used by many systems analysis and design tools, which enables simple multicode integration across simulation tools. Specifically, RAVEN provides the user with a set of APIs such that external simulation codes can be interfaced directly to RAVEN itself. From a modelling standpoint, RAVEN also provides solutions to create internal data communication flows between simulation models.

Methodology: Depending on the goal and the problem, RAVEN works by sampling the response of the system to parameter alterations. It applies these perturbations using one of a variety of techniques, such as Monte Carlo, Latin hypercube sampling or reliability surface search. Third party software models the system and RAVEN accesses the results directly through software coupling or indirectly via input and output files.

RAVEN then manages the analysis of the generated data by standard classic statistical techniques or through more advanced data mining approaches. Machine learning, for example, can ease the computational burden of analysis through the construction of a simpler software representation of a physical model. RAVEN runs internal checks to ensure that such a software construct is sufficiently accurate.

RAVEN also oversees the parallel dispatch of computational tasks. This parallelization is part of an approach that allows RAVEN to run on platforms ranging from laptops to high performance clusters.

---

[50] Repository: https://raven.inl.gov

3.4.31. SALOME

SALOME[51] is an OSS that provides a generic pre-processing and post-processing platform for numerical simulation. It is based on an open and flexible architecture made of reusable components.

SALOME is a cross-platform solution. It is distributed under the terms of the LGPL. The SALOME software platform is an open framework that can integrate scientific solvers for modelling various physical domains and perform a wide range of computing simulations. It was developed jointly by EDF and the French Alternative Energies and Atomic Energy Commission for the modelling of industrial equipment for nuclear power plants, considering the design of new generation reactors, nuclear fuel management, reliability and safety, and material ageing assessments.

SALOME can be used as a standalone application for the generation of CAD models, for meshing for numerical calculations and for post-processing. Its main features are interoperability between CAD modelling and computation software, integration of new components into heterogeneous systems for the numerical computation and multiphysics coupling between codes.

3.4.32. SCIANTIX

Bridging lower length scale calculations with the engineering scale simulations of fuel performance codes requires the development of dedicated intermediate scale codes. SCIANTIX[52], an open source 0-D standalone computer code [74–77], was designed to be included or coupled as a module in existing fuel performance codes.
Methodology: The models currently available in SCIANTIX cover intragranular and intergranular inert gas behaviour in UO_2, and high burnup structure formation.

3.4.33. SCONE

SCONE (Stochastic Calculator Of Neutron transport Equation)[53] [78–81] is a Monte Carlo particle transport tool designed to maximize flexibility and ease modification. It supports full neutron physics of interest in reactor problems and agrees well with a representative sample of the MCNP (Monte Carlo N-Particle transport) criticality test suite. It supports most of the common photon transport

---

routines and can perform transport with multigroup or continuous energy nuclear data. It has OpenMP parallelization.

Methodology: Standard Monte Carlo radiation transport. The modifiability and ease of use is due to the use of modern Fortran, with heavy use of its object oriented features.

3.4.34. MFront

MFront[54] [82, 83] is a code generator that translates a set of closely related domain specific languages into plain C++ on top of the libraries provided by the TFEL projects. These languages cover the following three kinds of material knowledge:

— Material properties (e.g. Young's modulus, thermal conductivity).
— Mechanical behaviours. Numerical performances of generated mechanical behaviours were given particular attention. Various benchmarks show that MFront implementations are competitive with native implementations available in the Cast3M, Code_Aster, Abaqus Standard and Cyrano3 solvers.
— Simple pointwise models, such as material swelling used in fuel performances.

MFront provides interfaces for many academic and commercial FEM, fast Fourier transform and material point method solvers.

Methodology: MFront is based on many techniques such as those for code generation and efficient linear algebra for small objects.

3.4.35. TALYS

TALYS[55] is a software package for the simulation of nuclear reactions, with an emphasis on nuclides with mass number 16 and heavier in the 1 keV–200 MeV energy range — thus for low energy neutron physics — but also other particles and higher energies. The package is used to evaluate nuclear data for the TENDL (TALYS-based evaluated nuclear data library) and JEFF (Joint Evaluated Fission and Fusion) nuclear data libraries. TALYS is also used to generate random ENDF files for the Total Monte Carlo (TMC) method (i.e. uncertainty propagation with random ENDF files).

Methodology: Deterministic; implementation of nuclear physics models.

---

[54] Repository: https://thelfer.github.io/tfel/web/index.html
[55] Repository: www.talys.eu

3.4.36. TrioCFD

TrioCFD[56] [84] is based on TRUST, a platform common to various applicative codes that use common TRUST tools, such as space discretization, time schemes, operators or information technology (IT) structures for HPC. TrioCFD focuses on specific physical models and tools for CFD such as the following:

— Turbulence (Reynolds averaged Navier–Stokes and large eddy simulation);
— Fluid–structure interaction (arbitrary Lagrangian–Eulerian approach);
— Thermal radiation (transparent or semitransparent media);
— Two phase flow (discontinuous front tracking, diffuse interface method and homogeneous equilibrium model);
— Sensibility analysis tool.

Development is ongoing and new updates are regularly released according to the needs of the TrioCFD users and their applications. Examples include the design of next generation nuclear reactors, upper plenum flow investigation and nuclear reactor vessel safety analyses. Other applications are also emerging, such as the modelling of salt caverns applied to hydrogen storage or steam separators. Methodology: Two space discretization methods are currently reliably implemented as follows:

— VDF: Finite difference volume method, and hybrid of the FVM and the finite difference method. Applicable on a structured mesh.
— VEF: Finite element volume method, and hybrid of the FVM and the FEM. Applicable on a triangular (2-D) or a tetrahedral (3-D) mesh.

The upcoming release will include the PolyMAC methods, applicable for any mesh type, even hybrid and non-conformal ones.

3.4.37. TRUST

TRUST[57] [85, 86] is an open source, HPC capable, thermal hydraulics simulation platform. It implements the general architecture and all the basic tools and methods necessary to solve numerical problems (e.g. diffusion and convection operators, standard Navier–Stokes, solvers, time integrators).

All these tools and methods allow TRUST to solve the thermal incompressible Navier–Stokes equations (using a conservative formulation) and,

---

more recently, any compressible multiphase problem. More importantly, these tools also constitute the building blocks for derived applications (among which TrioCFD is the most notable one) specialized for a given technological problem (such as next generation reactors, but also proton exchange membrane fuel cells and lithium ion batteries).

Methodology: TRUST offers two space discretization methods: VDF and VEF (see Section 3.4.36).

The upcoming release will include the PolyMAC methods, applicable for any mesh type, even hybrid and non-conformal ones.

3.5. SOFTWARE THAT IS NOT OPEN ACCESS

The following software is open source but does not yet provide open access. It is, however, available on request.

3.5.1. CORE SIM

CORE SIM [87] is used for the calculation of static neutron flux, of all eigenmodes and their adjoint functions and of the neutron noise in the frequency domain and its adjoint function, both for critical and subcritical systems. It provides a graphical user interface for visualizing the input and output in three dimensions and requires minimal input file writing. The code was developed in MATLAB and successfully tested in Octave.

Methodology: Two group diffusion theory, finite differences, frequency domain for the noise.

3.5.2. ExplosionDynamicsFoam

This tool[58] is aimed at all reactor safety relevant combustion regimes [88–90], such as H_2–CO–steam mixtures. It integrates previous developments of ddtFoam.

Methodology: FVM built on OpenFOAM.

3.5.3. High Temperature Reactor Code Package

The High Temperature Reactor (HTR) Code Package (HCP)[59] [91–93] allows the simulation of several safety related aspects of a HTR core in a highly

---

[58] Repository: https://gitlab.lrz.de/dario_zivkovic/explosionDynamicsFoam

[59] Repository: https://nucleus.iaea.org/sites/oncore/SitePages/Registered%20User%20Area.aspx (IAEA NUCLEUS account required).

integrated manner. The main objective of the code is the "Design, implementation and validation of software for the simulation of [(very) high temperature reactors] by applying the latest programming techniques and standards" [91]. HCP has full 3-D capabilities. This is not available in any other integrated HTR code system, especially not for pebble bed reactors. HCP replaces all the legacy codes developed at the Jülich Research Centre over more than 40 years from the 1970s until 2010 and has been transferred to the IAEA to continue development and distribution.

Methodology: Based on legacy codes (VSOP, ZUT/TUZ, TINTE, MUPO/TISPEC, MGT-3D, TOTMOS, DIREKT, FRESCO-I, FRESCO-II, PANAMA, NAKURE-99, ORIGEN-JÜL).

3.5.4. OpenFOAM_RCS

The OpenFOAM_RCS[60] [94] project aims at supporting maintenance by adopting a modern IT development environment, wherein institutions conducting research with OpenFOAM can integrate and share their software and simulation set-up developments in a sustainable manner. The project provides its partners with two GitLab repositories, one containing software and one containing simulation setups. The software repository includes the customized code and constitutes an add-on to the OpenFOAM release from the OpenFOAM Foundation. The repository is regularly updated and upgraded (on a three to six month cycle or on request) to keep track of the latest developments in OpenFOAM and to integrate novel functionality. This entails, for example, additional models for the simulation of polydisperse bubbly flows, an interface for coupling external material property libraries, or functionality for evaluating differences between simulation results and stored reference solutions. The add-on also features test, tutorial and validation cases.

Methodology: FVM.

3.5.5. Versatile Advanced Neutronics Code for GPU-Accelerated Reactor Designs

Applicable to PWRs with a rectilinear structure, Versatile Advanced Neutronics Code for GPU-Accelerated Reactor Designs (VANGARD) [95, 96] can perform multicycle depletion calculations with an inline thermal hydraulics solver.

Methodology: Pinwise simplified P_3 (SP_3) source expansion nodal method.

---

[60] Repository: https://gitlab.hzdr.de/openfoam/nuclear (access limited to project partners).

4. UNIQUE OPEN SOURCE CHALLENGES

4.1. OPENING THE SOURCE

For the majority of time that nuclear energy R&D has been taking place, software for nuclear energy applications was developed solely as closed source or proprietary software. However, over the past 15 years there has been remarkable growth in nuclear OSS, resulting in both new codes that have been released openly as well as legacy codes that have been released under open source licences after long being closed. The rapid growth of the OSS community has led to a number of unique challenges faced by software developers, maintainers, contributors and users in the field.

To understand the challenges faced by OSS developers in the nuclear community, stakeholders across a wide range of organizations — including national laboratories, universities, private companies and government — were asked to provide input.

This section summarizes the viewpoints from presentations at the IAEA Technical Meeting on the Development and Application of Open Source Modelling and Simulation Tools for Nuclear Reactors, held in June 2022, and discusses the experiences of the contributors. The views expressed here are an interpretation of the collected information and do not necessarily reflect the views of any one individual or organization. Although an attempt was made to consolidate viewpoints from a diverse set of perspectives in nuclear software development, undoubtedly there are other perspectives[61] that may differ from what is included here.

From the meetings with nuclear OSS stakeholders and a review of the literature, several main themes emerged with respect to unique challenges. These themes are community building; funding and incentive structures; measuring success; maintenance and quality issues with OSS; and licensing and distribution. Each of these topics is discussed in the following subsections.

4.2. COMMUNITY BUILDING

The communities that interact with software projects are often what defines the health, success and longevity of a project. Many aspects of building

---

[61] For example, the survey did not include individuals who see themselves as exclusively users of OSS in the nuclear community. The interviews primarily capture issues that developers prioritize. However, user challenges based on the experience of the co-authors and colleagues are also noted.

an OSS community are challenging, particularly in the nuclear energy field. These challenges arise from differences in stakeholder perspectives (such as where the software is developed and the needs of those developers) as well as from building and sustaining a thriving user community. Balancing the needs of various community stakeholders while maintaining code quality and user support and allocating time for stability and feature development is challenging, as all compete for limited developer time. Furthermore, there are complexities and challenges in how decisions are made for the project and in establishing a vision for future project development.

In the following subsections, the topic of community is discussed in more detail. The community referred to will shift depending on the context of the discussion. There are several subcommunities that may be referenced that comprise the larger software community, such as the following:

— Users. Users of the software of all skill levels. They may contribute feature requests and bug reports but they do not contribute code or new features.
— Contributors. Individuals who contribute code to the software and engage with the user community but may not be part of the core development team.
— Developers and maintainers. The individuals mostly responsible for the software and the day to day changes made to it. They fix bugs, create a roadmap for the project, may have leadership positions within the larger governance structure, create new features for the software and work to support the users and their questions.

Challenges in community building arise among and within each of these groups. What may be a challenge for one group may be a necessity for another; for example, answering dozens of user questions may be burdensome for a developer but important for the users. This subsection describes challenges in community building from the perspectives of these groups.

Building a community around a software project is dependent on the software's maturity, how widely used it is within the field and the level of interaction that the core development team is seeking from the community. Many projects start out because the developer has a research problem and writes software to solve that problem. In this case, the community for that project may be limited to a few users and a single developer. Alternatively, software may start out as a research project but gain attention within the field and start growing with respect to the number of users, developers and contributors. Community interactions (e.g. communicating between community groups, planning for the project) differ strongly in each case. This subsection focuses on how community building is impacted by project growth, project governance, the required education and training of community members, and institution.

4.2.1. Project growth

All software projects change over time. Different stages of a software project's lifetime have different needs to which the project needs to adapt and each stage has distinct challenges that affect the project. The following four stages of development are used here:

— Early development is characterized by a small number of individuals developing a tool in relative isolation. The project may have some users, but not many.
— Intermediate development is characterized by a similar number of individuals as in early development, but the project is gaining attention. The number of users is growing and new features are being added to the project. The API may shift but the project has matured past early development.
— Mature development is characterized by a large community of users and contributors to the project. There are fewer bugs than in the previous development stages and a relatively stable API. Development may add new features, but API changes and new features are carefully considered for their impact on the existing user community.
— Late stage development is characterized by code that has existed in the community for a long time and has recognition. Some features may be obsolete owing to changing infrastructure. Development has transitioned from adding new features to primarily keeping the code up to date with modern tools.

Two modes of growth are implied in this list. First, the growth of the code itself; the code matures and approaches a state of stability. Second, the growth of the community; users, contributors and developers as they join the project and grow the community. While the two modes are correlated with one another — as a code becomes more stable, the community of support grows — each mode of growth presents different challenges.

As a piece of software and its associated community grow over time, the effort required for peripheral activities (e.g. answering user questions, updating documentation, developing education and training materials) may become significant and often is not explicitly supported by funding sources [97]. As the project community grows, so too does the number of questions from users and contributors. These questions are important, as they can help to uncover bugs, improve code quality by revealing nuances in the code and encourage better documentation habits. However, the more questions there are, the less time can be spent on development or the R&D tasks that are the main responsibility of the developers. Developers therefore need to balance the demands of R&D work against the need to invest in the community. Failing to invest in the community can have adverse effects, as community members who feel left behind may choose not to use the code or may convince other users to adopt an alternative software. This can lead to future funding difficulties for the project as success metrics decrease over time. On the other hand, developers may not be seen as achieving success metrics in their job if they do not complete development and R&D tasks. Hiring a scientist to perform exclusively software engineering duties in a research institution is difficult, although not impossible [98].

Thus far, it has been considered that community interaction is handled by the developers of the project. In large open source projects such as NumPy[62], there are community managers who help to guide users and contributors in this space. Community managers require funding and, as they are not necessarily academic or research professionals, it can be difficult to fund such positions within research institutions. More broadly, the lack of investment in community support can be viewed as an undervaluing of care work in software [99]. This is a systemic issue and will change only with significant community advocacy for structural change to make this work visible.

Many developers choose to invest in the community in the hope of achieving some community sustainability. That is, by creating positive interactions in the community early on, they will recruit loyal, invested users who generate an interest in the code. Such users will be competent enough to answer other users' questions or even make contributions to the code themselves, lessening the pressure on the core development team. In practice, achieving this stability in the community is difficult, especially without incentive structures in place to encourage developers or users to participate in this way. Additionally, only users and contributors who have the time and resources to do so will be able to participate in open source projects in this manner. Without broader support in the field, this means that growth of the communities may be limited.

As a piece of software and its community grow over time, other indirect unsupported work can become significant, such as incorporating contributions from the broader community. Similar to answering user questions, addressing and incorporating code contributions from newcomers is a high value, high reward activity. Newcomers who contribute in this way may become future developers and ease the maintenance burden of existing developers. Conversely, not engaging with newcomers can stifle future project growth. Engaging with these contributions can be challenging for projects. If the contribution changes the code design or the API, it can affect existing users, so existing developers

---

[62] NumPy is a powerful open source Python library used for numerical computing. It provides support for large, multidimensional arrays and matrices, along with a collection of mathematical functions to operate on these arrays efficiently.

need to invest time (and increase their workload) to guide the newcomer to a contribution that will be accepted by the community. Advance planning, including project roadmaps, issue and feature request labelling and active communication channels[63] help to address some of these challenges and can reduce the number of changes requested on the newcomer's contribution. However, this can be difficult to execute in a decentralized community where users and developers have different development cultures and expectations.

OSS packages and libraries have grown in the nuclear field and, in parallel, so have the tools that support them. This creates a lower barrier to entry for newcomers to start participating in projects and helps to support decentralized community projects. Tools such as version control, version control hosting web sites like GitHub and GitLab, continuous integration services, package managers, packaging indexes, integrated development environments such as Visual Studio Code, and software forums help community members with diverse knowledge backgrounds to get involved in software. It also means that maintainers and core developers need to have expertise or some level of fluency in all these tools to effectively manage the OSS project and to engage with the community.

As a project grows and gains new features, it can be difficult to maintain performance. In nuclear OSS, performance is often critical, especially in tools that are used on projects with limited resources or require high performance. Developers and maintainers may wish to incorporate contributions from the community but cannot do so if the performance is reduced. Not incorporating new features can negatively impact the tool development community or shift functionality elsewhere. This balance is difficult to achieve for software projects.

Ensuring feature stability in software helps to build and maintain user trust. As the number of users increases over time, breaking changes in the code that disrupt users become increasingly consequential. Users will rely on every observable behaviour of an API[64], so needlessly changing it will cause breaking changes in user scripts. This forces developers to be more mindful of breaking changes as the code reaches later and more mature development stages.

Some projects address the anticipated challenges of technical debt by investing significant effort in the user facing code in advance. While this is quite rare in the nuclear software domain, it is done in OSS projects more broadly.

---

[63] Communities that have multiple mechanisms for community communication, such as discussion forums, Slack, Gitter and community meetings, tend to minimize unexpected contributions from newcomers by making it easy for newcomers to get involved in community discussions before submitting code contributions.

[64] This is known as Hyrum's Law.

Developers may spend time making sure that the software is designed to be internally modular and extensible to minimize future maintenance burdens. As a result, user facing changes are minimized over time, even if the back end of the code shifts. This requires significant upfront effort on the part of developers, which may not be accounted for in scientific research projects that have a short lifetime. However, this investment will help to make the science easier to reproduce. If the tools that performed the science have longevity, so does the science that it supported. Additionally, most nuclear software QA guidelines explicitly require upfront design and definition.

Anticipating a stable API and internal modularity early in a project can prevent future challenges. Neglecting these factors can lead to technical debt and hinder development progress. Addressing technical debt is difficult to justify in research, and retiring outdated tools can disrupt the community. How to encourage new projects that prioritize minimizing technical debt remains a challenge.[65]

4.2.2. Governance

Open source projects involve community driven decision making. Governance structures vary by project and institution, but they play a crucial role in project success. Early stage projects often lack formal governance, relying on individual leadership. As projects grow, formal governance becomes essential to prevent stagnation and facilitate community involvement. Decision making processes can range from individual authority to consensus based models. Sharing commit rights is a key aspect of open source development, but it can be influenced by institutional rules and project goals. Some projects prioritize community contributions while others maintain a more centralized approach. Balancing these factors is crucial for long term sustainability.

In the United States of America (USA), American Society of Mechanical Engineers (ASME) Nuclear Quality Assurance (NQA-1) [100] certification is attractive for commercial applications, so it is pursued by some mature projects. NQA-1 has specific requirements related to the traceability of code contributions and the developer's qualifications. As a result, maintainers of NQA-1 certified projects tend to be centralized at a single location or institution. While the code is open and accessible to all, the decision making lies with the maintainers, and much stronger restrictions are in place to gain write access to the code, if write access is given at all.

---

[65] How to solve this problem is not clear. How can the community be encouraged to start new projects in a way that minimizes technical debt even if existing structures do not value that investment? Is technical debt inevitable? How to decide when to retire a tool and start something new?

Another challenge to transitioning to a completely open governance structure, where any person can contribute and eventually become a maintainer, is that codes that are developed in isolation may appear to compete against one another. This is especially true at institutions that have historically had large software development presences internally, with sibling institutions having mirroring development groups, such as the United States Department of Energy (USDOE) national laboratory system. Making these codes open source has perceived risks, such as how one ensures that bad actors do not gain write access to the code; how the institution provides reliable jobs for existing developers if a competitor is doing the same thing; and how the unique features of the code do not spread to competitors. These questions may challenge, if not directly contradict, an open source approach for a piece of software. Sometimes these institutions also have a political motivation for keeping code closed or supporting compatibility with institution specific codes. Distributing project governance and responsibilities among multiple institutions in an OSS stack may require investment and coordination across institutional leadership, sponsors and the development team.

4.2.3. Development culture, skills and training

Many projects seek to grow their user, contributor and developer communities to become more sustainable. Sustainability in this context means having a thriving group of software users, contributors and core developers that ensure that the code remains performant, stable and useful. In the nuclear engineering community, the users, contributors and core developers of domain specific software draw from very small pools of individuals with domain knowledge in nuclear engineering as well as skills in software engineering and computer science.

Many software development practitioners involved in the interviews learned how to program without formal training in software development best practices. Their development skill set has been influenced by the research groups that they participated in or other software projects that they learned from. This creates a strong difference in skills and experience as students transition into professional roles and as professionals interact with other software groups. Getting people to adopt new practices different from what they first learned can be challenging. Positive changes are, however, occurring at universities with engineering programmes as they integrate software development into their curriculum, which helps students to develop practices that are more standardized. Even so, concerted training efforts and universities have the potential to make further improvements on this front.

Consider two extremes in community software. First, a single tool exists that performs a task, and no other project overlaps with it. All community members contribute to the tool and there is no competitor. Second, every developer works on their own tools and builds up a framework from the beginning. Neither extreme is beneficial for the community. In the former case, where people incrementally build on well recognized existing software tools, the number of people who truly understand the underlying components, algorithms and structure will decrease over time. In the latter case, individuals spend all their time developing tools to carry out their work without making use of existing solutions. Furthermore, a proliferation of duplicate packages will be confusing to new community members, and no single package can be reliably trusted without a solid testing user base and longevity in packaging. Ideally, there should be some balance between these two extremes wherein limited duplication in functionality serves a useful purpose: allowing code to code comparisons and validation and enabling individual projects to adapt to community needs, in cases where that balance is not obvious.

Many developers emphasized the benefits of developing software from the beginning. Building a project from the start helps developers to learn software infrastructure skills that would be hard to develop when working on an established project. For students, building a new piece of software is a way to build knowledge and expertise — both in software development and in the underlying physics — that would be difficult to acquire otherwise. Conversely, building a tool with functionality that already exists may be 'wasted' effort. Discovering that a tool is not necessary in an existing ecosystem can be demoralizing, as can be finding out that someone else has duplicated this tool.

Often, learning new skills for software is considered monodirectional; that is, there is a seasoned mentor teaching a student. In software training, this is not always the case. Students working in industry, research institutions or academia learn skills from their mentors and colleagues. However, established professionals often learn new skills from students too. Students are aware of emerging tools for development and are exposed to a large variety of software approaches in their curriculum. Students are also expected to continuously adapt to new tools. More established professionals (e.g. senior researchers, faculty members) may have made careers on a single tool or technology and may not have been exposed to other practices for a long time. It can be difficult to convince individuals in these roles to change, but when change is achieved at this level, it can be beneficial for all involved.

OSS products can additionally assist in training. Learners can see code and learn from seasoned developers as the software is being developed and learn by example with production level code. Creating pathways for students to be involved in open source projects early on in their education can have a major positive effect and help them to transition to professional software development roles after their final degree. However, if a project is difficult to learn, is not well documented, does not accept community contributions or has a development culture that is hostile to the student, it can be a significant negative setback and can even deter a student from continuing in software development altogether. It is important for mentors to guide students to appropriate open source projects in order to cultivate good experiences in open source development for training.

With any domain specific software, there is a requirement to have some working knowledge of the programming language, software design and the domain. Nuclear engineering involves the solution of complex problems. Systems can involve very different physics, requiring different solvers, and their behaviour can evolve over very different spatial and temporal scales. These problems may require significant computational resources to be relevant to real systems, and developers may need to manage scaling and memory requirements to obtain adequate solutions. Some projects are so complex that, in practice, users need to be experts themselves.

As software becomes more specific, more tailored to single problems, more performant or more complex, fewer people have the knowledge to meaningfully contribute to its development. Many nuclear codes rely on multiple programming languages, thus requiring developers to be sufficiently fluent in multiple computer languages and in inter-language code communications. Developers also need knowledge of software design and tools, such as build systems, parallelization, unit testing, continuous integration, packaging and distribution, and documentation. In many software projects in the community, developers need a working knowledge of computer science, software engineering and nuclear engineering, which can be a significant barrier. Ideally, multidisciplinary teams that include domain scientists, computer scientists, applied mathematicians and software engineers will work closely together via well defined requirements and interfaces. However, this ideal can be difficult to achieve, given the level of resources available and the inherent challenges of working across organizational boundaries.

As projects mature and reach a level of stability, they tend to be more difficult for new members to contribute to.[66] A stable, mature software project lends itself to fewer easy to implement features that are attractive for new contributors to address. New contributors will need skills in nuclear engineering, software engineering and computer science, in addition to having an intimate

---

[66] Potentially this means that projects are not suited to training student developers in perpetuity. Rather, the best time to learn from a project is as it grows over time and meaningful contributions require less upfront investment from the contributor.

working knowledge of the software, to contribute to the mature software project. This potentially creates a higher barrier to entry for new contributors.

A team may also wish to curate a development culture within a software project. Some software projects establish a code of conduct, which outlines the expected behaviour of community members. Other projects may rely on workplace culture to influence the software project. Because OSS development is predominantly on-line (and sometimes developers rarely meet in person), communication that is clear and respectful is highly valued. This is usually outside the scope of a high level document such as the governance document, although members higher up in the software governance may choose to establish some guidelines. Many tools have emerged in recent years that help with community communication (e.g. GitHub, Slack, Gitter, Google Groups) and make communication more accessible for community members. However, fostering a specific culture within a community takes time and effort, and the development community may not have the dedicated resources needed to achieve this.

Navigating different development cultures and different development expectations can be a significant challenge for newcomers. Onboarding newcomers is a high priority for software teams that aspire to grow their development or contributor communities. However, onboarding itself can be a challenge. Maintainers may publish formalized development practices to help newcomers to learn the community culture, but in practice it is often more casual: a lead maintainer may reach out to walk a new member through the community, or the newcomer may learn the code as part of a summer internship. The formalization of onboarding and of maintainer mentorship requires time on the part of community members. Mentorship in an internship role may be remunerated, but the creation of community documentation rarely is. Some maintainers choose to spend extra time onboarding newcomers in the hopes that those newcomers will eventually help to bring more people into the project community. This work is largely unpaid, unrecognized and unrewarded, even though it is important for the growth of a project.

4.2.4. Additional perspectives from academia and industry

Discussions during the technical meeting elicited several themes about the differences in software communities throughout the stages of maturity of a project. It was found that there were project to project and institution to institution differences in software development practices.

Academic institutions serve to educate and train the next generation of scientists in the nuclear field. This, by design, means that most academic positions are temporary and time limited. The result is that software developed at academic institutions often does not benefit from a stable, long term core development team, compared with software developed in industry or other research institutions, and more time needs to be spent onboarding new developers. If software developed at a university has one primary developer, it can be difficult to continue development when that person moves to a different institution. However, constant turnover also requires the adoption of good code documentation practices that ease developer transitions. Constant turnover of developers makes maintaining production level software and retaining development momentum at universities challenging. However, it also has the benefit that software contributions may be easier to incorporate, allowing for the onboarding of a broader community and contributor base to the software. It also reduces the risk that a single developer will exclusively retain all historical knowledge of the project. Universities are also where most developers involved in the interviews learned and developed their software skills. University software projects serve as an important training bed for new methods and a place for learning new software skills and working in the broader community.

Developing and contributing to OSS in industry has its own set of unique challenges. In industry, most projects have a clear profit motive and have to be supported by management. Depending on the management's receptiveness to OSS, additional barriers may be faced by developers. Developing software and releasing it to be freely available can be perceived as not leading to any revenue or, worse, giving away the company's competitive edge and intellectual property for free. The software in development has as an end goal to advance their product, technology or service without compromising future deliverables. This has been addressed by some companies offering consulting or direct developer resources in turn for financial support of the open software. However, even under such a funding model, it can be challenging internally to allocate resources to developer time when staff are busy with other projects that are seen as contributing more directly to the mission of the company. Nevertheless, the open source approach can be viewed as a positive way to engage in a public–private partnership by private companies.

Even with supportive management and leadership, developers can face difficulty in getting time to develop or contribute to OSS. Universities and national laboratories usually have defined procedures for creating open source tools, although these procedures may be onerous at times. This is not always the case in industry. Navigating logistical and procedural issues to develop software may be difficult or even prohibitive. Nevertheless, the industry is generally supportive of the use of open source tools. Even if it is difficult to release internal code with an open, permissive licence, OSS is usually relied on in the software stack for industry. Workers may be able to contribute upstream in the form of bug reports and fixes if these do not appear to reveal technology sensitive information.

4.3. FUNDING AND INCENTIVE STRUCTURES

As outlined in previous subsections, software developed in the nuclear energy field can be developed or can transition to OSS with the support of management, stakeholders and the community. These may be loosely considered as procedural or social barriers to OSS. Nevertheless, every team interviewed viewed funding and incentive structures to be among the most challenging aspects of OSS development in nuclear engineering.

The majority of nuclear open source scientific software is developed at research institutions and academic institutions. These institutions have specific R&D goals, and many researchers develop software tools to assist in their scientific pursuits. Most software in this space is developed in support of research and has funding until that project ends. Even if the software tool is developed at the inception of a funding award, awards are typically for only a few years, whereas it can take five years[67] or more before an R&D tool becomes a stable general use code or library. However, the software exists and can be used past the end of fiscal sponsorship. This means that developers either have to find a new scientific problem to which the tool can be applied (and extended) or they need to seek out funding to maintain the software elsewhere.[68] It can be extremely difficult to get sponsorship for projects at this point.

There are very few programmes that fund software directly. Funding opportunities that do exist for software are often limited in scope or targeted at a specific toolchain. It can be very difficult to get a tool funded if it exists outside an existing toolchain, especially if it overlaps with a previously funded tool. Moreover, it can be extremely difficult to add an existing tool to one of these toolchains, and it can take years of effort and persistence, which may not be possible for early career researchers. Funding software projects is often viewed as a zero-sum game; if funding is awarded to a new project, it may be viewed as taking money away from an existing, previously funded project. This creates tension within the software community and can make community building more challenging. Moreover, projects may be evaluated by sponsors on the basis of metrics that are not applicable across all projects. A new tool may show promise but have few users, while an established tool may have many users but few of the core team are doing R&D. This is discussed further in Section 4.4.

Funding opportunities do not always reward open source practices. For instance, within the evaluation criteria for the USDOE's Office of Nuclear Energy funding frameworks, there are no incentives for projects to use or produce OSS

---

[67] Five years is about how long it usually takes to reach this point (see also Section 5).

[68] Having a deprecation period where one promises to maintain a feature only for a set amount of time could help to incentivize further support from sponsors.

or to make analysis pipelines open. Merit reviewers do not have criteria to assess or reward the 'openness' of a tool or project. This may be changing, as there have been shifts within USDOE — both the Office of Nuclear Energy and the Office of Science — and the United States Nuclear Regulatory Commission that appreciate open artefacts. In the European Union, there is more top-down support for open source tools. European agencies now have a practice of assigning additional merit to proposals that use open source tools.

Funding for maintenance and user support is especially difficult to acquire. Although sponsors — such as academic institutions and governmental entities — value scientific R&D, maintenance and user support are often not considered part of R&D. Maintenance and user support are crucial to the success of the code;[69] software stability builds trust with users, who then recruit new users through positive experiences. Only one project discussed in the interview was able to get funding exclusively for maintenance and user support. Maintenance and user support are complex to measure, and sponsors may not see tangible outputs from these activities. Most projects fund both maintenance and user support through supporting projects[70]. This underscores the importance and value of communicating needs for maintenance and support to sponsors. Some institutions do have overhead support for software maintenance, but other laboratories do not have this avenue available to use for their software projects. Some developers feel pressure to continue maintenance and user support for a tool that they want to succeed, even if funds do not exist for them to do so. As a result, they perform these activities in their spare time. Although this benefits users, as it results in software stability, it ends up devaluing scientific developer time.

There are other funding avenues for software maintenance, such as direct support from stakeholders through a sponsoring agency like NumFOCUS[71] [101]. In the case of NumFOCUS, sponsorship awards can be directed to NumFOCUS with lower overheads. Overheads at research and academic institutions can be a major barrier to getting direct support for maintenance from non-governmental sponsors, as they make developer time more expensive, meaning that less support

---

[69] One could also argue that stable software helps to produce future scientific R&D more reliably.

[70] 'Supporting projects' here means projects that provide funding to the OSS either for consultancy or for fixing bugs and developing specific new features. Supporting projects can also be grants that fund science associated with user support or bug fix maintenance but are not included in the grant directly.

[71] NumFOCUS is a nonprofit agency that was founded to support maintenance of scientific software projects. Many tools in the scientific Python stack are fiscally sponsored projects under the NumFOCUS umbrella.

can be provided per unit of currency spent. This also contributes to the expense of the software development of the project itself.

Much of the preceding subsection on funding discusses groups or institutions seeking financial resources for software. Open, community governed codes are not tied to a single institution, however. Research institutions will often choose to invest in software that is of strategic benefit to them. An in-house code with good user support and unique capabilities helps to attain future funding for the institution because the institution houses the tool. Finding a way for open tools to thrive in this framework of funding and institutional support is challenging.

The use of community code means that the project can evolve and adapt to community needs, but it may also mean that no institution feels responsible for stewarding the code, for ensuring its success and for pursuing funding on its behalf. The effect is that a few very motivated individuals at institutions advocate on behalf of these tools to obtain organizational support through internal projects, institutional overhead[72] and others. Not every individual has the political capital to do this, which means that individuals in more vulnerable groups will have diminished access to these careers.

In academic and research institutions, scientific software developers are under pressure to produce outputs that are relevant to research metrics[73], such as research paper publications. Software products (e.g. distributions, releases) are not valued at the same level as research publications. This creates tension for developers in these institutions in two ways. First, scientific developers focused on creating stable, reliable software will spend time investing in that software and will produce fewer research publications, resulting in fewer rewards than those of their peers. Second, peers who want to succeed in these institutions will not value or invest in software stability, creating more tools that are unreliable and unstable. Both of these ways negatively impact scientific software progress over time. Industry colleagues may be hired according to software development relevant performance metrics, potentially allowing them to not face these challenges at the same level.

4.4. MEASURING SUCCESS

When developing any piece of software, open source or not, important questions that need to be considered by all stakeholders (i.e. developers, sponsors

---

[72] MOOSE and OpenMC are examples of community codes that have organizational support [66, 102].

[73] Section 4.4 describes software success metrics in further detail.

and the wider community) in the software are whether the software has been successful and how success is defined. The answers may vary depending on who is answering, where they are answering from, and what the goals of the software team are. There are also related questions such as whether or not tracking specific, quantifiable metrics is necessary.

One of the most obvious measures of success for any piece of software is the size of the user community; that is, how many people are using the software. For CSS packages in the nuclear energy field, the easiest way to measure use is by looking at the number of licensed users. This method does have pitfalls, namely, that it is difficult to ascertain whether a user who obtained a licence years ago is still actively using the software. However, it still gives a good sense of the use of the software over time. In the open source model, users are not required to register or explicitly fill out a licence form, so it is not possible to know how many people are using the software at any given point in time. There is a wide variety of other metrics that can provide a measure of the software's use and activity. These include the number of downloads/clones from a version controlled repository[74], the number of people who have starred or favourited a repository, the number of active users and posts on a user forum, the number of active contributors and the number of contributions over time. One has to be careful not to put too much weight into these simple metrics, as any one given metric can be easily manipulated. For example, if the number of GitHub pull requests[75] were used as a measure of community activity, developers could be incentivized to submit many small pull requests to boost the numbers.

The above metrics give an indication of the raw size of the user community, but because so much of the activity in the nuclear energy community is focused on R&D, it is also important to understand the level of use of a software package specifically for research activities. For this purpose, arguably the best single metric is the total number of times that a software package is cited in journal or conference papers and the number of citations over time. Citing software in publications presupposes that there is one or more reference papers or technical reports that users can cite, which is not always the case. Although some software developers have published reference papers, not all developers have the time or willingness to do so. There currently are several mechanisms for software developers to receive

---

[74] Downloads ought to be carefully considered as a metric, as continuous integration services often download software in testing runs. This has the potential to inflate the downloads metric unless it is accounted for.

[75] In other software hosting services, a request to merge a change is referred to as a 'merge request' or a 'change request'.

proper attribution in citations without necessarily having to write a traditional journal or conference paper. Several journals specifically publish papers about software, including the following:

— Journal of Open Source Software;
— SoftwareX;
— Journal of Open Research Software;
— Computer Physics Communications.

These journals also allow authors to focus on aspects of the software design and engineering that might not be suitable for discussion in a more general journal.

Software citations provide additional context that simple metrics such as GitHub stars or clones miss, such as where a software package is being used and for which applications. Success of software is often judged not only by how large the user community is but more so by who is using the software. If a software package is adopted by an influential nuclear reactor designer, startup company or regulator, this is seen as more of a success than having many users that are less well known. Use of software by students at universities is also viewed as having a positive impact, as it helps to develop a talent pipeline for prospective employers. The breadth of use by different organizations is difficult to quantify; nevertheless, it serves as useful anecdotal evidence that software developers can present to their sponsors and point to when trying to secure funding.

In addition to software oriented journals, software developers can take advantage of software archiving services such as Zenodo[76] and FigShare[77] that automatically assign a digital object identifier (DOI) to releases or uploads of a software package. The DOIs associated with software releases provide users with an effective means of citing a specific version of a software package, which aids reproducibility. For software developers funded under the USDOE, the software services platform DOE CODE provides another means of collaboration, archiving and discovery for OSS. In conjunction with these services, software developers can include a CITATION.cff file in their repository, which lets others know how to correctly cite the software and which is supported by GitHub and Zenodo.

---

[76] Zenodo (https://zenodo.org/) is a research data repository and platform that allows researchers to share and preserve their research outputs, including datasets, software, papers and other research artefacts.

[77] FigShare (https://figshare.com/) is an on-line platform for sharing research outputs and data. It allows researchers to upload and publish various types of research output, including datasets, figures, posters, papers, presentations and other scholarly materials.

Ultimately, even when developers make it clear how to cite their software, it is still up to researchers to include proper citations for software in their authored works. All too often, authors do not properly cite software that was integral to their results, which makes it harder for the software developers to demonstrate impact. It can be difficult for authors to know when software needs to be cited. For example, if a researcher writes a script that relies on several third party Python packages, it is unclear which of these packages (if any) are deserving of a citation.

In the nuclear energy field, software developers often have a wide variety of roles and responsibilities beyond just developing software. This can include grant writing; project management; supervision of staff, students and postdoctoral researchers; teaching and training; and engagement in collaborations. For most research positions, the number of publications and impact of publications are still key metrics for performance evaluation and career advancement. Some of the citation mechanisms mentioned previously (e.g. software oriented journals, software release DOIs) may not be as widely recognized or valued as traditional journal publications, which could be viewed as a disincentive against using these mechanisms.

Thus far, the discussion has been focused on the success of software, with an implicit assumption that one of the goals for software developers is to build a user community. However, in many cases, software is developed primarily to serve the needs of its original authors in achieving their research or industrial or commercial goals, and release of the software under an open source licence may be driven not by their desire to build a user community but simply by funding or sponsor requirements, convenience (e.g. simplified collaborations, access to free services) or other considerations. In this case, the question of what constitutes success becomes more complicated. Clearly, the number of users, downloads and software citations would not show strong evidence that the software was successful, but to the original authors, achieving research goals may be the only meaningful measure of success. This underscores the importance of viewing software success in terms of the original goal of the software. Sponsors too may ultimately view success in a similar manner, primarily focusing on what is accomplished with the software rather than how general the software is, how large the user and developer community is or how often it is cited.

Software is often one of the most important outputs of research, and there is strong evidence that building a community around a piece of software can help to improve the software's longevity, maintainability and overall impact towards advancing science. For government funded research, developing a community for an OSS package is also highly effective at providing a strong return on investment for taxpayer money while ensuring the widest reach and impact.

4.5. MAINTENANCE AND QUALITY ISSUES WITH OPEN SOURCE CODE

There is a concern about the maintenance of OSS produced in a university or academic context as opposed to commercial proposals. For example, the CANDU Owners Group in Canada has the requirement that a production code used in refurbished CANDU reactors needs to be supported in the period 2030–2060. Such a requirement is possible only with the application of rigorous QA processes. Open source code needs to fulfil such a requirement to be considered.

Production code requires rigorous QA processes, including source versioning and revision, issue tracking, continuous integration and availability from an official repository such as those proposed by the Nuclear Energy Agency (NEA) Data Bank. Here, the term 'source' includes code, non-regression tests, documentation, configuration scripts and issue tracking information.

There is a misconception about the fact that QA techniques are taught in universities and used in the industry. If OSS is produced in a university, QA techniques need to be applied for its development. It would be highly beneficial for the community if the OSS used by the nuclear industry were hosted in a dedicated location, ideally by an independent international body.

4.6. LICENSING AND DISTRIBUTION

In principle, releasing OSS is simple: as a software developer, once a point is reached where the software is ready to be released, it is simply uploaded to a source code hosting service such as GitLab, GitHub or Bitbucket and it is then available to the world. However, there are many things that need to be considered when releasing the software: What licence should be assigned? How will users install the software? What level of support will the original developer's team provide to prospective users and developers? Who maintains control of the software when contributions come from multiple organizations? Who owns the software? What does ownership even entail? These are all questions that researchers may be ill-equipped to handle, and it can be difficult to find reliable sources of information.

When software is intended to be used for nuclear energy applications, a different set of considerations apply. Is the software subject to export control regulations? Are the research and its products (including software) considered fundamental or basic research? Navigating these questions is especially difficult and, in the discussion with software teams, it was one of the main areas of confusion.

While some of these questions may be deferred until they arise, others need to be addressed as part of the software release. The first question is regarding which licence to choose. Developers have dozens of options, and the differences between many licences can be very subtle. The OSI maintains a list of 'approved' licences, and at present there are over 100 approved licences to choose from. They also provide a list of nine licences that are popular. For nuclear energy related software, perhaps the most important question is whether to use a copyleft licence or a permissive licence. Copyleft licences (e.g. GPL) allow derivative works but require them to be released under the same terms as the original work. This means that if an organization makes a modification to the software, they are required to release it under the same licence. Permissive licences (e.g. the BSD licences) have no such restriction, so derivative works can be released under any terms (including as part of a larger commercial product). In the discussion with software developers at private companies, it was clear that industry has a strong preference for permissive licences, as they do not need to consider any legal implications for including such software in a larger product. In the survey conducted for this publication, projects were approximately evenly split between using permissive licences (e.g. BSD, MIT/X) and copyleft licences (e.g. GPL, European Union Public Licence (EUPL)).

Once a licence has been chosen, developers also need to consider how their software will be installed by users (i.e. how the software will be packaged and distributed). Packaging and distribution of software are among the more complex and rapidly changing areas of software development. At the simple end of the spectrum, developers may choose to release only a source package, and the burden is then on the user to compile/install the program. However, for ease of use, they may wish to package and distribute the software using whatever distribution mechanism exists for the programming language in which they are working. While the software has certainly been designed to work on the system used by the developer, issues commonly arise when trying to install or use the software on a different platform. This is an area where software developers need to be proactive, as ensuring that software can work across multiple platforms and compilers requires a concerted and non-trivial effort.

To highlight the complexity of packaging, the following describe the considerations for two of the most popular programming languages for scientific computing, C++ and Python. Starting with C++, some of the issues that developers deal with include the following:

— How to ensure that the software can be built across multiple platforms. Several solutions for cross-platform builds exist (e.g. CMake, autoconf, Meson), and while CMake is increasingly becoming the predominant choice, there are still some widely used software packages that do not rely on CMake (e.g. MOOSE).

— How to build against third party dependencies (e.g. include the software as a 'vendored' dependency, require that a user first installs the software from source, rely on a package management system).
— Which distribution channels should be targeted (e.g. Spack, vcpkg, Conan).

In Python, the following equally challenging set of considerations comes into play:

— Which packaging system to support (e.g. PyPI, Conda, Linux system package managers);
— How to handle distribution when the Python code calls for underlying C/C++ libraries;
— How to handle breaking changes in dependencies.

4.6.1. Export control

In the discussion with software teams, one topic that came up in almost every discussion was the challenge of dealing with export control laws and regulations. On the one hand, it can be difficult to address the topic of export control in a general manner, given that each country has its own set of export control regulations and these may differ from one another. However, a group of 48 countries have entered an international agreement through the Nuclear Suppliers Group (NSG) to coordinate export control practices. The NSG has two sets of guidelines that govern the exports of nuclear related technology, INFCIRC/254 Part 1 [103] and INFCIRC/254 Part 2 [104], which together are known as the NSG Guidelines. Consequently, the export control practices in individual NSG countries are usually aligned with these international agreements. As one example, much of the language of the export control regulations for the USA closely follows that which is found in the NSG Guidelines.

Many software teams work with both OSS as well as CSS that is subject to export controls. Therefore, some general areas of difficulty may not be directly related to open source development but are challenges that most software developers in the nuclear energy field are likely to encounter. Three common themes emerged: (a) understanding fundamental research exemptions, (b) interpretation and application of export control regulations, and (c) categorization of software under different export control systems, particularly in the USA.

The NSG Guidelines make a specific exception for basic scientific research, which is defined as experimental or theoretical work undertaken principally to acquire new knowledge of the fundamental principles of phenomena and observable facts, and which is not primarily directed towards a specific practical aim or objective. In the USA, an analogous concept originated from a 1985 policy directive, NSDD 189 [105], defining fundamental research as "basic and applied research in science and engineering, the results of which ordinarily are published and shared broadly within the scientific community" and outlining the policy that products of fundamental research should remain unrestricted. This definition of fundamental research appears in all US export control regulations [106, 107].

While the fundamental research exemption is very broad, there is considerable confusion over what qualifies as fundamental research and whether a given piece of software will be exempt from export control under this exemption. This creates uncertainty for software developers, as a review for export control is typically done at the point where the software is being released instead of prior to its development.

In the USA, the overall picture is complicated by the fact that there are at least four organizations that have some jurisdiction over export control regulations: the Department of Commerce, USDOE, the Department of State and the United States Nuclear Regulatory Commission. Nuclear related software usually falls under USDOE regulations and the Department of Commerce, but often teams are confused as to why a particular piece of software is under one or the other. There are several instances where multiple software packages having similar functionality were categorized under different regulations, as well as instances where one piece of software was export controlled and another with similar functionality was open source.

In addition to the lack of clarity around export control jurisdiction, there is also considerable confusion over the written language in the regulations themselves. For example, under the Export Administration Regulations Commerce Control List, software for neutronics calculation/modelling or for radiation transport calculations/modelling is subject to control. However, no further definition of 'neutronics' or 'radiation transport' is given, leaving organizations and their export control officers to interpret these terms. Although these issues are, at face value, not relevant to OSS, several software teams that were involved in the interview have a desire to make their software, or some subset of their software that is currently under export control restrictions, open source.

Export control restrictions on software have created challenges at university nuclear engineering programmes, which rely heavily on software for training and education purposes. Export controlled software has become increasingly difficult to use in academic settings because of the lead time necessary to secure licences, and because changes to software made for R&D purposes cannot easily be transferred from one person to another. These challenges weaken the talent pipeline for organizations that serve as prospective employees for new graduates. One concern expressed by teams that manage export controlled software is that the difficulty in obtaining licences at universities has made it harder for employers to find students with experience using their software.

4.6.2. Monolithic and modular software

In thinking about the future of OSS in the nuclear community, there is a major question around whether large software projects should be designed as monolithic packages[78], which are all in one packages with very few external dependencies, or modular packages, wherein each unit in the software stack has a specific purpose and a well defined API and is typically managed as its own project. This question is closely intertwined with the question of how to distribute software, since the two models may have very different distribution mechanisms.

The observation from encounters with many teams is that there is a lot of momentum towards the modular software model. As the nuclear OSS ecosystem grows, it is important to recognize the pros and cons of these two approaches to have a good understanding of what can be gained and what is lost by moving in one direction or the other. This understanding also serves to highlight some of the challenges that software teams are likely to experience if they are redesigning a monolithic software package into modular components.

Historically, most software packages used by the nuclear community were designed as monolithic applications. These codes are characterized by large codebases, limited or no third party dependencies and the use of a single programming language. Many of these codes have existed for decades and were designed in such a manner out of necessity. While it may not be immediately obvious, there are some important advantages to this approach. First, the lack of third party dependencies can significantly simplify software development. In this model, developers do not need to worry about upstream dependencies changing their API or dropping support for a particular platform or compiler. The monolithic approach also simplifies licensing considerations because there is no need to consider any licence terms for dependencies; the entire monolithic application can be packaged and distributed under whatever licence the developers choose. The lack of third party dependencies can also simplify software QA requirements. Many organizations prefer to have a monolithic application in order to attain the scientific and technical goals for a project. When changes are needed in some component of the software, they have full control and are not reliant on an upstream package accepting a merge or pull request.

The main downside of monolithic applications is that the developers are responsible for the burden of maintaining every component in their software.

---

[78] The terms 'monolithic' and 'modular' may take on different meanings in different contexts. For example, monolithic software (an all in one package with no external dependencies) may itself be internally structured in a modular fashion. In the present discussion, the difference relates to whether a given piece of software is designed (modular) or not designed (monolithic) to interact with other pieces of software.

This often includes low level numerical algorithms and solvers, input–output and communication libraries, and other basic functionalities. When new issues arise in such components (e.g. they do not work with a new compiler), the developers have to address them rather than relying on a wider community to do so. Developers also do not benefit from continual improvements that would typically be made in a community maintained library.

The ability to create new applications while building on existing work is one of the key drivers of the modular software approach. As the complexity of scientific computing applications increases over time, it becomes increasingly difficult to maintain a single monolithic application that can serve all needs. This is particularly true with the growth of multiphysics applications, where two or more codes that solve for disparate physics are coupled to one another. When software developers present a well defined interface that external tools or codes can consume, it becomes possible to build more complex applications without reinventing every piece of the software stack. As alluded to above, modular software can significantly reduce the maintenance challenges associated with large software projects because the complexity is distributed across many independent components. Relying on well maintained community packages also enables software teams to take advantage of improvements in those underlying packages. This is especially true in scientific computing, where there is often a need to take advantage of cutting edge computing architectures and optimized parallel algorithms.

One of the major challenges for modular software is packaging. Managing the packaging and distribution of modular software is certainly more challenging than it is for monolithic applications, but as packaging tools and environments have improved, it has become easier to maintain a stack of multiple software components with many interdependencies. For example, in Python, dependencies can be easily expressed in packages released on PyPI and conda-forge. For C, the Spack package manager has made it possible to build complex software stacks on a wide variety of architectures, including clusters and supercomputers. While there are still many issues and opportunities for improvement, the availability of these packaging solutions has significantly improved the experience for scientific software developers.

For nuclear software, modular software design is an important tool for ensuring that components of software that are not sensitive can be made open source. For example, if the geometry routines from a large — and otherwise export controlled — codebase were isolated into a library with a well defined API, they could be released under an open source licence while still being used as part of the larger codebase. This approach can also lead to better stewardship of taxpayer money by ensuring that the products of publicly funded research are made available to the maximum extent possible. A good example in the nuclear OSS community is MOOSE [102]; the parallel finite element framework itself is open source, while some applications that are built on top of it are under export control restrictions.

5. LESSONS LEARNED AND BEST PRACTICES IN NUCLEAR OPEN SOURCE PROJECTS

5.1. OVERVIEW

A few decades of developments in various institutions have produced several lessons on open source projects, with the possibility of deriving some general best practices that take into account the specificities of the nuclear industry (see Section 5.2). This section discusses some best practices related to the integration process (Section 5.3), community management (Section 5.4), QA (Section 5.5), documentation (Section 5.6), programming (Section 5.7) and intellectual property management (Section 5.8).

Best practices depend on the field of modelling; on the objective of the developed software; on the type and size of the organization that develops the software; on whether the development is driven by an institution or by a community; on whether the software is developed from the beginning or from existing software; and on whether the software is designed to interface with other software. The specific field of modelling (e.g. CFD versus neutronics) strongly impacts the size of the user community; the availability of validation cases; the size and complexity of the software; and the possibility to benefit from developments outside of the nuclear community.

The objective for which software is developed strongly impacts all development requirements. As an example, software that is developed for reactor licensing purposes will have increased requirements in terms of QA and V&V compared with software dedicated to research, education and training. Conversely, software for education and training, in particularly, will benefit from transparent, high level programming. In addition, software may be developed mainly for internal use or as a source of revenue via software licensing and/or consulting services, which strongly affects decisions related to software licensing and documentation.

The type of organization that develops the software will determine the possibility and effectiveness of implementing certain best practices. Typically, a company will employ a more stable workforce compared with a university, with somewhat more limited turnover. As another example, universities and research centres often lack mechanisms to hire and/or assign personnel for non-research oriented tasks (e.g. documentation, licensing, distribution). Compared with institution driven projects, community driven projects will normally benefit from a larger developer base, but will also face a somewhat slower integration process and more challenging QA. Finally, software that is developed from the beginning provides full freedom in terms of programming best practices, whereas existing software may limit choices somewhat (e.g. the programming language, the programming paradigm).

Although a detailed discussion of best practices for specific situations is beyond the scope of this publication, some guidelines that generally apply to open source nuclear software development are proposed. Where relevant, the most important implications associated with the modelling field, objective, type of organization, type of governance and initial development strategy are also discussed.

5.2. OPEN SOURCE PROJECTS IN THE NUCLEAR COMMUNITY

Nuclear organizations are faced the need to understand and model complex physics for the design, operation and maintenance of nuclear plants. Part of these needs can be addressed using general purpose methodologies and codes (i.e. general CFD or FEM codes for structural mechanics). However, for complex phenomena specific to the nuclear industry, such as neutronics, two phase flow in fuel bundles and the behaviour of materials under irradiation, specific simulation tools have to be developed when commercial solutions are not available. For over 50 years, a suite of codes in the areas of thermal hydraulics, structural mechanics, electromagnetics, free surface flows, neutronics and fuel behaviour, for example, have been developed by various institutions in order to address specific modelling and simulation challenges. For some codes, the developers have decided to follow the open source approach.

The following specificities of the nuclear industry have been observed to significantly affect the way OSS is developed in this field:

— QA requirements. The specific nature of the nuclear technology often entails the need for more stringent QA protocols compared with other industrial fields. As an example, in the USA, ASME proposes the NQA-1[79] certification programme, which is endorsed by the US Nuclear Regulatory Commission and USDOE as providing an acceptable basis for complying

---

[79] For more information about the NQA-1 programme, see https://www.asme.org/certification-accreditation/nuclear-quality-assurance-nqa1-certification

with their requirements. Achieving NQA-1 certification entails a full audit of the QA programme performed by trained ASME auditors.

— Potential dual use and export restrictions. Nuclear technology, including software, is typically subject to export restrictions. This may pose significant limitations to an open source development approach, notably when the developed software requires industrial data for calibration.

— Size of the community. Software employed in the nuclear industry often addresses modelling needs that are specific to the industry, which drastically reduces the potential number of users and developers.

— Validation data. Depending on the specific field of modelling, data for software validation are often limited, restricted or very expensive to produce.

— Existing software. Nuclear industries and regulators sometimes rely on in-house tools that feature relatively old modelling approaches and numerical algorithms, but that benefit from decades of validation and calibration on proprietary data. This combines with the hurdle associated with QA and validation of new software to make these industries resistant to rapid changes and fluid software development.

— Specific modelling requirements. The need to model unique phenomena (e.g. radiation transport, two phase flow in reactor cores, fuel behaviour) makes inheriting work from other fields difficult, although the situation is changing with the gradual adoption of general purpose numerical libraries and the increased use of machine learning.

5.3. NEW DEVELOPMENTS AND INTEGRATION OF NEW SOFTWARE

The integration of various developments in open source projects can adopt standard pathways when these projects are driven by single institutions that handle both decision making and the actual integration process. In this section, it is assumed that the open source project has some level of involvement of the community.

5.3.1. Decision making

A crucial aspect of OSS development is associated with managing the contributions from the community. The way that this is done strongly depends on the modelling field, the objective, the type of organization, the initial development strategy and in particular the type of governance.

When the development is driven by a single institution, or by a group of institutions via a steering committee, decisions are ultimately taken by the institution or steering committee itself. However, successful open source projects have highlighted the importance of the following:

(a) Communicating development directions in advance. This is important both to allow for informed decisions from the community on whether or not to invest resources in the use and/or development of the software and to limit the risk of the community duplicating work that is already planned.

(b) Communicating development needs and criteria of acceptance for community contributions. This is essential to maximize the benefits of an open source development strategy, while avoiding the risk of the community working on valuable contributions that cannot be integrated into the main project (e.g. because of a different programming style).

(c) Involving the community in the decision making process. This is crucial for the success of an open source project. Although individuals or organizations may decide to adopt an OSS for economic reasons, or to support education and training, the long term involvement and active contribution of the community require mechanisms to involve the community in the decision making process. For small user communities, the most important development directions and decisions may be published in advance and the community may be allowed to comment on them via specific channels (e.g. emails or Internet forums). In the case of larger user communities, this involvement can translate, in its simplest form, into regular polls prepared by the governance and proposed to the whole community. Finally, for very large projects, experience has shown that direct inclusion of the community in the governance structure can be effective. This may be achieved by creating technical committees dedicated to specific areas of development, as well as special interest groups focusing on specific areas of application.

Best practices in the decision making process for projects driven by an organization can be summarized as follows:

— Indicate the development directions.
— Indicate the development needs and criteria of acceptance of community contribution.
— Involve the community in the decision making process.

In cases where the development is driven by the community, defining the rules for accepting new contributions is crucial. One possible model consists in accepting contributions approved by at least one developer. In turn, one can become a developer by having at least one major contribution accepted in the code. In the case of disagreements, a small steering committee of major developers can intervene to settle the dispute. This model is proving effective for a relatively small project such as OpenMC. Considering that most nuclear related projects tend to have a relatively small user base, adopting a similar model for community driven projects may be beneficial. Variations may be introduced, such as the need for approvals from multiple developers (e.g. at least two) and the need for these developers to belong to different institutions (to avoid a single institution's taking the lead in the project). In this case, it would also be useful to publish clear guidelines for code contribution to help the community to provide usable pieces of software.

5.3.2. Integration process

Integration of new software from a heterogeneous set of developers requires a systematic strategy and rigorous implementation in order to provide a certain level of QA, guarantee backward traceability of developments and allow for a rapid de-implementation of features and code modifications. These requirements essentially translate into both the adoption of a software version control system such as Git and a continuous integration strategy with automatic testing.

Version control systems allow tracking and control over changes to a source code. They keep track of modifications to the source, allowing developers to compare earlier versions, for example to spot a mistake or to repeat old calculations. This guarantees backward traceability and repeatability of previous results. Version control systems also allow different developers to work on new features or fix bugs while minimizing disruption to other team members. Finally, it is common practice to maintain two main versions of a code: a production version that includes only stable code and an advanced version that includes new developments and that is meant for merging into the production version after a period of testing. Among the various available code versioning systems, Git currently seems to be the most common choice in the nuclear field.

From a software user perspective, version control systems add the ability to include a fully defined version indicator of the software used, even when it is between formal releases. This facilitates the formal documentation of input files, output files and code version, ensuring traceability between analysis results and any subsequent reactor design decisions.

Version control repository hosts often include the ability to define and enforce certain rules regarding who is allowed to change different branches of the software and under what conditions. Such features readily facilitate typical nuclear software QA requirements, including the ability to maintain a list of qualified software developers and to require and document the review and acceptance of any proposed changes.

Use of a version control system has been observed to yield the best results when associated with an automation system that monitors the repository for changes and automatically triggers a series of checks on any proposed change. Checks can include unit testing, integration testing, user interface testing, accessibility testing, code format checking (linting), testing coverage measurement, performance testing and other examinations. By automating these checks, a project can establish rules that provide automatic feedback to developers and that need to be met before engaging a qualified code reviewer to evaluate the change. This enforces project standards, improves consistency, reduces human error and increases the efficiency of code reviewers. Importantly, interpersonal conflict can even be reduced by keeping trivial discussions out of code reviews, such as commentary about whitespace and formatting.

Such a version control system is often associated with continuous integration, an agile best practice where developers integrate their code changes into the main code version early and often. The goal is to reduce the risk of complex integration at the end of a project, with version control that allows for a well ordered and traceable development even with a multitude of continuous contributions. The main risk of continuous integration is associated with cumulative errors that may make it impossible to fix bugs and retrieve a functioning version of the software. For this reason, continuous integration is ideally accompanied by automatic regression testing, where code contributions are accepted in the main code version only after passing automated tests that prove that they do not affect the other code functionalities, or by specific testing designed to prove that new developments affect results only in the way intended by the code verification process (e.g. via comparison with analytic results, via the method of manufactured solutions).

For organizations that maintain and develop both software and databases, close integration of software and data management by applying the aforementioned continuous integration strategy with automatic testing based on available databases would be favourable. For example, adopting periodic regression tests involving a large set of data and simulation cases may be considered. This would greatly improve data fairness while providing an additional layer in the QA protocols, allowing for easy detection and backtracking of anomalies in software or data. With time, this would also provide invaluable statistics on the evolution of quantities of interest predicted by different combinations of data and codes.

Best practices in the integration process can be summarized as follows:

— Use of a code versioning system;
— Full automation of testing and linting;
— Adoption of a continuous integration strategy with automatic testing;
— Integration of software and data management.

5.4. COMMUNITY MANAGEMENT

Although it is difficult to estimate the time needed to build an active community of users and developers, a useful guideline is to consider a period of ten years, including the following:

— Five years to prepare the software and let people know about it. In this phase, it has to be expected that there will be a need to provide significant assistance to users and new developers, with limited feedback, few declarations of interest from institutions and some interest from individuals.
— Five years to consolidate the software and expand the community. In this phase, the software is more mature and better known, which can trigger academic and industrial partnerships and a growing community that can support both the development and the V&V of the code.

During this ten year period, different strategies can be employed for community management that depend heavily on the size of the user and developer community and on the development status. Projects in their earliest development phases featuring small development communities and even smaller user communities normally benefit from the use of private discussion forums (e.g. messaging programs) where one can interact with other users and developers directly or in topic specific channels. These forums have also been observed to remain important tools in later stages of development, allowing effective interaction among groups of developers.

Mailing lists can be another important tool in the early stages of software development for triggering initial community growth and engagement. However, a mailing list can become difficult to manage, as the list of recipients can rapidly evolve into a very heterogeneous group (e.g. previous and current developers, commercial partners, users) with different expectations in terms of communication. For this reason, it is proposed that their use be restricted to the very early phases of a project.

Public forums normally manifest their benefits later on in projects. First, they allow large communities to be provided with important information. Second, public question and answer (Q&A) session forums significantly relieve the developers of user support duties. With time, such forums tend to become essential complements to the documentation. In this sense, it can be beneficial to open a public forum early in the process and to populate it with Q&As from various sources, including the private forum and personal communications.

Training courses, including massive open on-line courses, tend to open the tool to a larger, potentially less experienced and more demanding public. They are essential for the success of an open source project; however, it would be most beneficial to delay them until the software is stable, with limited expected changes in the input deck and functionalities. Course participants invest significant time and effort into learning, and developers ought to avoid challenging these efforts with rapid changes in the code input deck or user interface.

Workshops and user meetings are common in nuclear industry conferences and meetings. They provide good avenues for code developers to expose a specialized interest group to their code and its capabilities. It is best to take advantage of the open source nature of codes in advance and either send attendees instructions on how to install the tool on their own computer or set up an on-line environment (e.g. JupyterHub) that they can readily activate once they arrive. The first impression is essential to engage potential users and developers, so the installation or getting started instructions need to be thoroughly debugged and tested.

Finally, the importance of an issue tracking system in an open source project is not negligible. Together with a discussion forum, an issue tracking system often represents the main communication channel between the users and the developers, and the only one that relates to the quality and usability of the software. If a user encounters a problem but has no easy way to report it to the developers for resolving, the user will most likely abandon the use of the software. More than a forum, the issue tracking system requires a timely and supportive response. Beyond supporting users, issue tracking systems facilitate the ongoing formal QA of software, which universally requires software issues to be identified, fixed, documented and linked to impacts on the software's performance. Systems that allow linkages between issues, source code changes and the resulting test suite changes may be used for such purposes.

Best practices in community management can be summarized as follows:

— Expect the need for a period of ten years to build a community.
— Set up an effective private forum as early as possible to enhance communication among developers and early users.
— Set up a public Q&A forum as early as possible. If needed, artificially populate it with questions and answers from various sources.
— Implement an issue tracking system and consider it an essential part of community management, not just a tool for QA.
— Organize training courses, but only in the advanced stages of software development.

5.5. QUALITY ASSURANCE

Software quality standards vary by region, but the following is normally expected in a software QA programme:

— Define the requirements of the software (what does the software do?).

— Describe how the requirements are implemented (how does it do it?).

— Demonstrate through tests how each requirement is satisfied (verification).

— Demonstrate that the requirements solve the original problem set out to be solved (validation).

— Define the bug reporting process and the ability of users to receive notifications defining new bugs and their impacts.

— Define the change management process governing what, who, how, when and where with respect to code updates.

In nuclear engineering simulation codes, verification can often be accomplished largely through an automated unit and integration test suite, while validation often involves benchmark comparisons with measured physical data of a relevant physical system.

All users of codes expect a certain level of quality when choosing to use specific software. If the code is inaccurate, unstable or fragile, users may feel that their time is better spent with a different approach. Institutional users who work under a formal QA programme will have much higher expectations and desires for the software's pedigree. The more the developer team can do to demonstrate high quality, the less the users will have to do to comply with their rigorous internal QA procedures while using the code for analysis of nuclear equipment .

Software developer teams can choose whether or not to attempt to become qualified software vendors under certain standards. If they choose to become qualified, they can offer pre-verified software to users, indicating that the code complies with a certain set of standards (e.g. NQA-1 2015[80]). This qualification generally requires a complete set of requirements and verification tests, a complete validation test suite and a major commitment to providing bug reports and fixes to users going forward, sometimes many years into the future. In some cases, the software vendor would need to agree to be subjected to regular audits from each customer's QA department and, potentially, the nuclear regulator. None of the open source vendors identified so far offer pre-verified OSS services. However, it was noted that such services could provide financial sustenance to a developer team that does not use the traditional 'software licence for fee' structure.

In open source projects, the following best practices can help a developer team to make the overall QA process most effective:

— Use of a revision control system is essential for a well ordered development strategy, backward traceability and repeatability of results.

---

[80] More information about NQA-1 is provided at https://www.asme.org/certification-accreditation/nuclear-quality-assurance-nqa1-certification

— Automatic testing avoids introducing significant mistakes and guarantees the consistency of results over time.
— Integration of software and data management enables more effective testing and provides statistics that can help to identify potential discrepancies over time as a result of improved data or code, or of newly introduced bugs.
— Use of an issue tracking system is essential to unlock the full potential of OSS in terms of QA, namely continuous testing and feedback from the community.
— Embedding and automatic generation of documentation avoids inconsistencies between code and documentation.
— Adherence to programming best practices minimizes the risk of introducing bugs and favour a correct understanding of the code for new developers. In an open source context, they are also essential to involve the community in the QA of the code.

In addition, it is essential to define a bug resolution protocol and communicate it to the community. A typical example of bug resolution protocol using Git is the following:

— Write a test that reproduces the problem.
— Create a new branch.
— Fix the code.
— Check that the test is now passed.
— Commit the code fix (and only the code fix).
— Use the cherry-pick command of Git to merge the commit into other code branches.

It is important to define the required qualifications of people who change the code and those who review the code and integrate it into production releases.

Open source projects are often characterized by two opposing mechanisms. On the one hand, an OSS holds high potential for deep V&V in that the broader community can more easily contribute to the process. On the other hand, it can be difficult to make the process consistent and well targeted. For example, it is not uncommon that the community will focus its efforts on specific domains. In an academic setting, these may be areas that are the subject of active research that can result in impactful publications. For an organization developing an open source tool, filling the V&V gaps left by the community can be extremely resource intensive. This is particularly the case for software intended to support licensing of nuclear hardware.

Open source projects ought to be structured to capitalize on the support of the community. Some users and developers will be genuinely interested in supporting validation efforts for specific applications, and it is essential to clearly communicate what the validation needs are and to define channels for the community to contribute validation cases, including mechanisms to acknowledge these contributions and promote the work done. It is also important to make V&V cases part of the continuous integration process as early as possible during the development of the code. Validation cases will quickly lose their value if they can only be performed on old code versions.

One of the major potential benefits of OSS in the nuclear domain is collaboration on the limited physical validation cases. Many of these data were originally generated in nuclear facilities that no longer exist and would be extremely difficult and expensive to reproduce or supplement. Some international efforts to collect and standardize benchmark data, such as the International Reactor Physics Evaluation Project Handbook [108], have provided an invaluable service. Many institutions build their own computer models of these same experiments repeatedly. A significant amount of relatively commoditized effort could be saved if more detailed software models (e.g. computer readable) based on these benchmark publications could be developed and shared. Then, when a transcription error in the historical data is identified by one group (often through discussions with retired technicians who remember the details), it can be readily shared for the good of the industry. Additional work is needed to define the best way to make general computerized models of these benchmarks that can be used judiciously by numerous software developer teams. The work by the Committee on Computer Code Coordination [109] done in the 1960s is still relevant today but could be updated to support more unstructured grids.

Best practices in terms of QA can be summarized as follows:

— Follow the best practices in terms of code integration, community managements, documentation and programming.
— Define a clear bug resolution protocol and communicate it to the community.
— Clearly communicate validation needs.Define channels for the community to contribute validation cases.
— Define mechanisms to properly acknowledge and promote the contributions from the community.
— Include the V&V cases in the continuous integration strategy.

5.6. DOCUMENTATION

High quality documentation is necessary for the success of an OSS and represents an important showcase for promoting the company or institution and the domain of application of the tool. Successful practices in documenting a code often involve preparing the following:

— A user manual, to describe the use of a code;
— A theory manual, to describe the theory and limitations associated with the code;
— An implementation or developer manual, to describe the implementation and guide developers;
— QA documentation, including a V&V report, formally defining software requirements and/or critical characteristics and demonstrating their testing (this is particularly important in the nuclear field, especially if the code is developed for licensing activities).

Providing an OSS with such documentation presents different challenges depending on whether the development of the code is led by a commercial company, or by research or academic institutions and communities. For example, a commercial company has to allocate a significant budget to this activity, while disposing of limited revenues with respect to development and distribution of licensed software. In addition, the task of writing documentation is often regarded as unrewarding by employees. Additionally, academic and research institutions often lack proper mechanisms to allocate resources to code documentation, which essentially has to rely on the goodwill of developers. This can make creating complete and consistent documentation a considerable challenge. However, some best practices can significantly help in the task.

Documentation for open source codes globally spans a spectrum, as follows:

— No documentation is provided, and the user has to read source code;
— Some documentation embedded in source code as comments and/or docstrings;
— Documentation published occasionally as one or more documents with different releases of the code;
— Documentation embedded in the same repository as the source code and generated or published with each update of the main branch via continuous integration.

The final option has arguably been the most convenient and successful. Modern software documentation build systems 'compile' the documentation from source text into documents in HTML, PDF and/or docx formats. The source text is often written in a kind of simplified plain text markup language that works seamlessly with software revision control systems.

For example, one could use an object oriented programming style based on C++ and include in the header file of each class a brief description of both its theory and its usage. One may then use Doxygen[81] to automatically generate documentation in HTML format. Another extremely powerful tool is Sphinx[82], originally created to document Python projects by combining narrative documentation written in a markup language (reStructuredText) with additional detailed documentation scraped and inferred from the docstrings of source code modules, classes and functions.

The approach of integrating the documentation and the source code together into the revision control system and using a documentation building system yields the following advantages:

— It reduces context switching by allowing developers to write and update documentation in the same environment and with the same tools as they are updating the code.
— It promotes the concurrent development of software and documentation, minimizing inconsistencies (e.g. during find–replace refactoring).
— It promotes the development of documentation that is performed by the code developers themselves, which, because of their expertise, may improve the quality of both documentation and code.
— It limits the work associated with formatting and styling of the documentation, since this is typically an automatic process performed by a selected tool.
— It allows rich and low maintenance hyperlinking between narrative documentation sections and specific implementation details down to the function level.
— It allows deployment of documentation to happen at the same time as deployment of new versions of the code.
— It allows the same documentation to be rendered in HTML format for user convenience as well as in PDF or docx formats to support any potential institutional quality record requirements.
— It integrates numerous manual sections into one single piece of documentation.

Experienced developers often struggle to imagine the mindset of new users and may skip over the proper documentation of important elements that have become familiar to them. If available, obtaining reviews and updates from skilled technical writers is invaluable. In the open source world, technical writing is another potential area where non-programmers can make meaningful and important contributions to a project.

---

[81] https://www.doxygen.nl/
[82] https://www.sphinx-doc.org/en/master/

5.6.1. User manual

The user manual describes to users how to install and use the code. The style and approach of this documentation is flexible, but successful user manuals typically contain some combination of a quick start, annotated tutorials and examples, and detailed narrative descriptions of the key user operations of the code.

Installation instructions often feature a quick start section that aims to guide the user from a blank slate to a fully installed code and a working simple example in 30 minutes or less. Quick starts allow experienced users to get going quickly and often link to more detailed or nuanced installation instructions covering less typical installation scenarios on another page.

Following the installation instructions, it is important to include a description of where users may go to find assistance. This could include links to mailing lists, public forums, chat channels and telephone numbers. Public forums and chats are an important complement to documentation. The creation, early in the project, of a discussion forum can help to generate a large collection of recurring Q&As in the long term.

Next, it is beneficial to include a collection of several annotated tutorials, possibly covering all the functionalities of a code. This minimizes the barrier to approaching the code and provides a good starting point for users, who do not need to recreate complex cases from the beginning. Annotated inputs can also be linked to the above mentioned automatic documentation. In academic settings, tutorials can be associated with published papers that can provide more details of the case.

Similarly, collections of gallery-like examples that demonstrate simple but useful functionalities provide users with quick working examples to help them to understand what the code can do as well as how to get the code to do it. Combining interesting figures or graphics with the code that was used to generate them is particularly engaging for users. Even experienced users may find themselves returning to a well designed gallery to perform common operations.

All tutorials and gallery examples are ideally built as a step in the continuous integration process. If a code change impacts the tutorials, the tutorials need to be updated before the code change goes into production. Otherwise, it has been observed that the code and tutorials diverge and users become frustrated when the tutorials lead to errors or unexpected behaviour.

Given a working installation and several successful simple runs, users will then need detailed instructions covering all the other user operations. This documentation needs to explain all possible model definition and approximation options and all runtime options and needs to describe the output data in detail. This section will generally be the largest and most detailed part of the user manual.

5.6.2. Theory manual

The theory manual generally defines the mathematical models that are implemented in the code. It may provide some background on the methodologies employed, define the governing equations and discuss the solution techniques employed to solve the equations given the user input. Bidirectional references and hyperlinks between the theory manual and the implementation or developer manual are extremely valuable to users and developers alike. Writing documentation with a system that supports equations is essential for the theory manual and such capabilities are widely available.

5.6.3. Implementation or developer manual

The implementation or developer manual generally explains how the code is designed and implemented to satisfy its objectives. It serves the following two key purposes:

— To show new or existing developers how to maintain, upgrade or interface with the code;
— To formally describe how each requirement is implemented in code for QA purposes.

If a documentation build system is used, much of the detailed implementation documentation can reside in the header files or docstrings of the code. The accessibility of the implementation documentation helps developers to update the documentation while the code is being updated.

Documentation (and even code comments) can become a maintenance burden. If an implementation is documented excessively, it will often become out of date with the actual code. For this reason, highly experienced developers often structure their source code and variable names to be self-descriptive, with only sparse explanatory comments. The trade-off between completeness and this burden is non-trivial to manage. Thus, documentation review has to become an integral part of the code review checklist in the code development process.

Another advantage of using documentation build systems such as Doxygen and Sphinx is that they can actually read through the structure of a code and automatically generate a consistent API definition showing all classes and methods, including their arguments and often the argument type. This amount of detail can become overwhelming for users, but it is useful as a reference to be hyperlinked to other more detailed parts of the documentation. A useful pattern is to describe something in narrative form and then provide a hyperlink to the API document section where the up to date details may be seen.

Tools such as Doxygen and Sphinx can generate class diagrams through inference of the code itself. Such automation avoids the need to update Unified Modelling Language (UML) diagrams after refactoring the code.

Finally, the source code ought to be considered an integral part of the documentation. A well written and commented code may limit the need for documentation and will greatly reduce the load on developers in terms of Q&As, while encouraging users to become contributors. In addition, one of the unique advantages of OSS lies in the possibility for users to investigate the code, which promotes a better understanding and a more proficient use of the software compared with closed source codes. However, for this potential to be realized, developers need to put effort into making the source code readable, well ordered and sufficiently commented.

5.6.4. Quality assurance documentation

The QA information discussed in Section 5.5 can be integrated into the main documentation. The exact location of the QA elements in a project's documentation is flexible, but some general guidelines can be applied.

Formal software QA efforts benefit greatly from the use of a software requirements database. These requirements track traceability between parent and child requirements, linkages to test cases and revision control of the requirements themselves (the change of which often has to undergo formal review by a panel of stakeholders). Thousands of commercial requirements management databases with these capabilities are available. For open source teams with limited funding, projects like the Sphinx-Needs plug-in are one example of a fully capable open source requirements traceability system.

When software is released as a tagged version, all the current requirements ought to be included, including a report of which requirements were met in working tests that were successfully completed during the release process. Attempts to quantify how much progress a developer team has made in fully qualifying their code include some basic measure of how many of the total requirements have been defined, reviewed and approved by the stakeholder panel, and how many of the approved requirements have been fully verified by automated tests.

Publications are often used in open source projects to supplement or replace the theory manual and QA documentation. Although this practice can significantly limit the need for dedicated documentation in academic settings, it is important to provide users with an annotated list of these publications to indicate what can be found in each one of them as well as what information is out of date.

Best practices on documentation can be summarized as follows:

— Include in the documentation a user manual, a theory manual, an implementation or developer manual and QA documentation.
— Embed the documentation into the code itself in a format that can be used for automated generation.
— Prepare a commented set of tutorials that cover the essential functionalities of the software.
— Prepare gallery-like examples that demonstrate simple but useful functionalities and provide users with quick working examples.
— Tutorials and gallery examples ought to be built as part of the continuous integration process.
— Create a public forum for Q&As early in the project.
— Prepare commented lists of publications that can serve as theory and V&V manuals.
— Use bidirectional references and hyperlinks between the theory manual and the implementation manual.
— Consider the source code an integral part of the documentation.
— Seek reviews and updates from skilled technical writers.
— Include information on where to find assistance.

5.7. PROGRAMMING

The selection of a programming language and paradigm (e.g. functional or object oriented, static versus dynamic typing, compiled versus interpreted) is a complex task and it depends on a wide range of factors including the size of the project, the need to interface it with available libraries and software, and the competence available in house. While it is beyond the scope of this publication to provide guidelines for the selection of a programming language, the community working on nuclear software tends to be quite small and to adhere to well defined trends. For example, there is a clear trend towards writing new nuclear codes in C/C++ (e.g. Serpent, MOOSE, OpenFOAM, OpenMC) and Python, which is consistent with a more general trend in the scientific computing community. C/C++ and Python can arguably be considered the infrastructure of much of the modern computing world. Concurrently, a large portion of nuclear legacy codes, and a small number of recent ones, are written in Fortran. Finally, among various domain specific languages, MATLAB is widely used, and some recent trends towards the use of the programming language Julia were also observed. When selecting a programming language and defining paradigms and coding styles, it is important to consider current trends in the community. Although this should not hinder innovation, the selection of a specific programming language and paradigm will largely determine the talent pool available to support the project, both as potential hires and as contributors and/or collaborators. For example, OpenMC recently transitioned from Fortran to C++, partly to access a wider pool of talents.

With regard to the programming paradigm, object oriented programming has features that are particularly useful for open source development. Object oriented programming allows the creation of fully encapsulated pieces of code (named classes) that can easily be shared with other users and embedded in different codes or solvers. As an example, OpenFOAM arguably owes part of its success to an advance use of the object oriented programming paradigm, which allowed for an extendable code with streamlined contributions from the community. A similar paradigm was more recently adopted in the nuclear field by the MOOSE environment, which has allowed for an extendable framework for rapid developments and resulted in a large set of solvers and significant involvement of the community.

Large amounts of software are being written by people without formal software development training, such as nuclear and mechanical engineers who picked up software development as part of a project in school and found continued use for their skills in their careers. Certainly, some true software experts did not come through a computer science programme, and with the prevalence of information work in the modern engineering world, large numbers of non-expert software developers are tasked with software development work. Pairing nuclear domain experts with seasoned software professionals has shown to be a successful pathway to move beyond merely functional code to maintainable/sustainable code.

The value of type hinting and checking is particularly relevant for an interpreted language such as Python. The value of such a consideration is reflected in the many large software companies that have created competing type-checking systems that can post-process Python code looking for typing issues. When a compiler is foregone in a statically typed language that refuses to build if, for example, an integer is passed to a function that expects a string, a large class of programming errors begin to show up only during runtime, ideally during testing, but often in practice during use. Type checkers run through the code, identifying these bugs before the code ships. While Python is excellent in its ability to foster developers from novice to productive skills, making large complex systems without the guard rails of type checking is a true hazard.

After selection of programming language and paradigm, the corresponding best practices need to be adhered to. Programming best practices are often recommended by developer organizations or can be learned from the experience of reputable institutions[83].

Best practices on programming can be summarized as follows:

— When selecting a programming language, take into consideration the current trends in the nuclear community. This will largely determine the talent pool available for the project.
— Consider adopting an object oriented paradigm to streamline collaboration and simplify code maintenance.
— Adhere to programming best practices as recommended internally or by other reputable institutions.

5.8. INTELLECTUAL PROPERTY MANAGEMENT

In OSS the most important aspect to consider is whether or not to adopt copyleft types of licence such as the GPL. Such licences impose conditions that the freedoms granted by the licence (e.g. to access the source code, modify and redistribute without limitations) apply to any derivative work. There are prominent examples of software licensed under the GPL, including, in particular, the Linux kernel. Copyleft licences have their pros and cons. On the one hand, a GPL licence will promote a truly open source development paradigm, where all developers will have a clear incentive to publicly redistribute their derivative work. On the other hand, it will be impossible, for example, to distribute code that dynamically links a GPL licensed code with a closed source code.[84] Stemming from this limitation, the Free Software Foundation has released the LGPL. The LGPL has mainly been created for software libraries and allows dynamic linking of an LGPL library with other codes and distribution of the resulting software without affecting its licence. More permissive software licences also exist that carry only minimal restrictions on how the software can be used, modified and redistributed. Examples include the GNU All-permissive License, MIT License and the BSD licences. Selection of a licence is a complex task that involves considerations related to institution or company policies, the business model, the need to use copyleft libraries and the need to link with proprietary software, and other considerations. However, the choice of a licence carries important consequences, and developers need to be careful in their choice.

---

[83] For instance, see https://ocw.mit.edu/courses/6-s096-effective-programming-in-c-and-c-january-iap-2014/; https://www.epfl.ch/labs/ivrl/teaching/teaching-projects/teaching-projects-requirements/recommended-coding-practices-cplusplus/

[84] While it is possible to do this linking and use the resulting code without distributing it, the GPL licence requires all derivative work to be distributed under the GPL licence.

Another important aspect of intellectual property management in open source projects is associated with 'forking' (i.e. the possibility that another organization would start developing a separate version of the software). Forking can be a sign of a healthy and successful development of a software. For example, a specific organization may decide to fork a project in order to specialize it in a specific domain that is not necessarily of interest for the entire user community. However, multiple forking can rapidly create confusion in the community and dilute community efforts. In addition, there is always the risk of ill-intended forking that is meant to replace the original distribution.

Two main mechanisms exist to limit forking and provide a competitive edge to the original version of a code and its derivative work. The first one is copyright. In most countries, and according to the Berne Convention [110], copyright protection is obtained automatically without the need for registration or other formalities. Copyright includes economic rights, which are normally waved by the authors from the moment they opt for an open source licence, and moral rights. Moral rights include, in particular, the right to claim authorship. This translates into the possibility of claiming authorship of each piece of written code. That authorship will be inalienable and can be used to protect the original authors of software. The second possibility to protect one own's work is the use of a trademark for the software's name, which allows one to regulate the rights of other developers in using it.

Best practices in terms of intellectual property management can be summarized as follows:

— Carefully select the type of open source licence, notably with respect to copyleft characteristics.
— Be rigorous in copyrighting the software before distribution.
— Consider protecting valuable software with a trademark.

6. OPEN SOURCE SOFTWARE FOR EDUCATION AND TRAINING

6.1. LEARN, USE AND MODEL

Use of software in education and training can have the following three main learning objectives:

— Learning to use a modelling software;
— Investigating physical phenomena using a modelling software;
— Learning modelling algorithms and gaining modelling skills.

6.2. LEARNING TO USE MODELLING SOFTWARE

Learning to use software as part of the educational and training process is associated with the benefits of modelling and simulation. Simulation allows the exploration of 'what if' questions and scenarios without having to experiment on the system itself. It facilitates the identification of bottlenecks in material, information and product flows, and provides insight into which variables are the most important to system performance. It provides an important method of analysis that is easily verified, communicated and understood. Across industries and disciplines, modelling and simulation provide valuable solutions by giving clear insights into complex systems.

In this context, CSS can present some advantages, as it is more likely to be a stable, focused product with well developed documentation. It can then reduce learning time and facilitate access to learners with more limited technical skills. In addition, well developed customer service often allows issues that may arise during training to be quickly resolved.

However, CSS often costs money, with costs varying from a few thousand to a few hundred thousand US dollars. In some cases, student or academic licences can greatly reduce the cost for universities but may still pose an organizational burden in that sufficient licences need to be purchased, and the university needs to guarantee that software is used by students (and staff) within the limits of the licence.

Another specific advantage of OSS is that that it has been observed to motivate students, since the students feel that they can 'own' the code. In addition, motivated students usually appreciate the possibility of accessing the source code for a better understanding of its capabilities and limitations. Training on open source tools also offers a more transferable skill, in that students will be able to make use of the tool in any organization they work for or for personal projects.

6.3. INVESTIGATING PHYSICAL PHENOMENA USING
MODELLING SOFTWARE

Modelling software can be used to investigate a given physical phenomenon or a specific problem or situation. In such a case, the software is merely a tool to generate results, from which an understanding of physical phenomena can be obtained. How a code is built is therefore of lesser importance. Moreover, knowing how to use a code might not be crucial, as pre-written input files or decks might be provided to the students to ease the task.

Applying simulations in education and training is important for several reasons, the most important of which is probably that learners will be engaged in active exploration and learning. This is an approach advocated in modern instructional and learning theories. Creating hypothetical realities or changing time scales in simulations might sustain this learning approach. Additional reasons for using simulations are a motivational aspect and the possibility of creating situations that are unacceptable in real life for reasons of danger, costs or time.

Also in this case, the availability of high quality documentation is often mentioned as the main strength of closed source tools. The numerical methods implemented often guarantee the convergence of the runs under a wide range of conditions. This allows the user to concentrate on the essence of the exercises, rather than on fixing input writing or tuning parameters to get the calculations successfully completed. Furthermore, OSS often requires compilation of the source code on the machine used by the students. This is a far from trivial task, especially when other libraries and tools are necessary. The installation and compilation of those codes represent, in some cases, a significant overhead. This problem could be addressed by using a remote sever or docker containers, but this also can represent a significant complication.

Once again, a main drawback of CSS is associated with acquiring licences and enforcing the proper use of the software. In addition, such tools are often delivered as executables only, which results in a lack of transparency. In the case of a run returning a result that cannot be understood on physical grounds, being able to look at the source code would allow better understanding of how the solution was computed. Nevertheless, it needs to be mentioned that such a task (i.e. tracking the computation of the solution in the source code) is far from trivial, especially if no manual on the methods and their implementation is available.

An important advantage of the use of OSS for teaching and/or training purposes is that it facilitates the sharing and the reuse of solvers, routines and exercises among different institutions and among students, without the need to start a model from the beginning, but instead focusing on the specific topic of a module or lecture. It also allows the student to continue practising with the tool after the completion of a course. Finally, it significantly lowers the entry barrier for nuclear newcomer countries.

6.4. LEARNING MODELLING ALGORITHMS AND MODELLING SKILLS

Learning the modelling algorithms used in nuclear reactor modelling represents a more complex endeavour than learning to use a modelling software or using it to investigate a given physical phenomenon. Although there are a few textbooks and even computer manuals describing the algorithms for modelling nuclear reactor systems, fully comprehending the techniques generally requires implementing them in practical situations. Clearly, this is an area of education and training where OSS provides unique capabilities.

As there is no licensing issue with OSS, students can directly use the techniques they develop or learn after the completion of the course or training. OSS also makes it possible to go deeper into the implementation details, such as parallelization, mesh handling and manipulation, and other modern numerical techniques. However, the learning curve associated with using OSS can be a significant barrier when considering, for example, university curricula, making training sessions on OSS more suitable for intensive and highly specialized courses, such as summer schools for PhD students. There is also the risk that the focus of the course or training session could switch from learning numerical techniques to learning a specific tool.

7. SUMMARY

It is clear from the breadth and number of organizations and countries contributing to this report that the topic is timely and resonates with the nuclear community of software developers. A number of issues and topics that impact their work were raised, including the following:

— The need for open source tools in the nuclear field;
— Unique challenges for open source tools and codes in the nuclear field;
— Currently available tools and projects;
— Lessons learned and best practices;
— Open source software for education.

What is clear from the meetings that were held and the data in this publication is that the open source approach adds value for the developers of tools in the nuclear sector as follows:

— The IAEA's ONCORE portal is seen as a valuable way of educating and coordinating the community around a core nucleus of common tools.
— The motivation for taking an open source approach in the nuclear field is the same as in any other field, with the exception that there are more restrictive controls on the software than in other fields.
— Open science funded by public money allows better sharing of data, development and better outcomes.

— Community building with OSS is both difficult and a critical aspect of success; with the addition of the nuclear sector, this adds a further barrier to overcome.
— Having access to OSS for training is important because, among other things, it promotes collaboration and sharing in order to advance innovation and it may significantly lower the entry barrier for nuclear newcomer countries.

Growth of OSS in the field of nuclear energy R&D is a result of new codes being openly released and legacy codes being made open source. Despite the many benefits for which the developers turn to OSS, there are still many challenges. The challenges faced by software developers, maintainers, contributors and users in the field include community building, funding and incentive structures, measuring success, and licensing and distribution. The challenges associated with community building arise from differences in stakeholder perspectives, the need to build and sustain a thriving user community, and to balance the needs of various community stakeholders while maintaining code quality and user support and allocating time for stability and feature development. There are also challenges in project growth, project governance, and the required education and training of community members.

The FAIR initiative promotes guidelines and best practices for data management to ensure that data are 'findable, accessible, interoperable, and reusable'. Open source software and the Linux operating system are examples of the collaborative approach to software development that has become synonymous with the open source movement. Open patents are a relatively new concept that aims to promote innovation by making valuable intellectual property available to a wider range of stakeholders. There are several examples of open design initiatives, including the Open Design Alliance and the Open Design Foundation.

Growing the community around OSS projects was highlighted as important, specifically how projects change at different stages of their life. Four stages of software development were identified: early, intermediate, mature and late-stage development. Growth of the code and the community present different challenges. As a project community grows, the effort required for peripheral activities — such as answering user questions and updating documentation and education and training materials on the code — may become significant. It is important to balance the demands of R&D work against the need to invest in the community, as not investing in the community can have adverse effects, including future funding difficulties for the project.

A project's governance structure can have an impact on the community. As a project grows and gains users, it becomes necessary to have a formal governance structure to ensure that development can progress over time. However, who makes the decisions for the project and how conflict is resolved will change over time and vary by institution and project maturity. Some mature projects in the USA seek NQA-1 certification, which requires centralized maintainers and specific requirements related to the traceability of code contributions. There are also challenges of transitioning to a completely open governance structure, particularly at institutions that have historically had large software development presences internally. While the various communities — users, contributors and developers — need to grow, this can be difficult, particularly in the nuclear engineering community, which draws from small pools of individuals with domain knowledge in nuclear engineering and software engineering.

Software development practices can vary between academic, research institutions and industry in the nuclear engineering field. Academic institutions often lack a stable, long term core development team, but constant turnover forces good documentation practices and may make software contributions easier to incorporate. In industry, there may be management barriers to developing and releasing OSS because of a perceived lack of a profit motive or fear of losing a competitive edge. Funding and incentive structures are among the most challenging aspects of OSS development in nuclear engineering. Most nuclear open source scientific software is developed at research and academic institutions, and funding for software projects is often viewed as a zero-sum game, which can create tension in the software community.

There are different metrics for measuring the success of software, particularly OSS, in the nuclear energy field. One of the most obvious measures of success is the size of the user community, but there are also other metrics that can be looked at, such as the number of downloads, active users and contributors. The level of use of software specifically for research activities can be measured by the total number of times that it is cited in journal or conference papers. The success of software is often judged not just by how large the user community is but more so by who is using the software. Software developers can take advantage of software archiving services, as well as software oriented journals and software release DOIs, which aid reproducibility and provide users with an effective means of citing a specific version of a software package.

Export control regulations were highlighted as a particular challenge in the nuclear energy field. There is confusion over what research qualifies as fundamental research and whether a piece of software will be exempt from export control under this exemption. There is also confusion over the language in the regulations themselves. The future of OSS in nuclear engineering was discussed, particularly software architecture; that is, whether large software projects should be designed as monolithic packages or modular packages. There is momentum towards the modular software model, but this approach presents its own challenges.

Best practices vary depending on the modelling field, the development objective, the size and type of organization, and the software's interface with other programs. Specific challenges faced in the nuclear industry include QA requirements, potential dual use and export restrictions, limited validation data and reliance on in-house tools. There are specific guidelines for OSS development in the nuclear industry, including community management, documentation, QA, programming and intellectual property management.

The integration of new software from different developers requires a systematic strategy and rigorous implementation to ensure QA, backward traceability of developments, and quick de-implementation of features and code modifications. Git is the most common code versioning system used in the nuclear field. Continuous integration is an agile best practice that integrates developers' code changes into the main code version early and often, reducing the risk of complex integration at the end of a project. Community management is crucial for the success of the software development project. Different strategies can be employed for community management, depending on the size of the user and developer community and the development status. Private discussion forums, mailing lists and public forums are some of the tools that can be used for community management at different stages of the project.

Software quality standards for nuclear engineering simulation codes are expected to meet specific requirements for V&V, bug reporting and change management processes. Software developer teams can choose to become qualified software vendors under certain standards, which requires adherence to a complete set of requirements and verification tests, a complete validation test suite, and a commitment to provide bug reports and fixes to users. Open source projects can be effective in QA through the use of best practices, such as automatic testing, integration of software and data management, and programming best practices. Open source projects ought to be structured to capitalize on the support of the community for validation efforts and to incorporate V&V cases into the continuous integration process. Finally, this publication notes the potential benefits of OSS in the nuclear domain for collaboration on limited physical validation cases and suggests that more detailed software models could save significant effort in reproducing experiments.

The choice of a software licence is crucial in open source development, and copyleft licences such as the GPL promote a truly open source paradigm, while permissive licences such as the MIT licence carry fewer restrictions. Forking, or the possibility of separate organizations developing different versions of the software, can be healthy or detrimental to a project. Copyright protection and the use of a trademark for the software's name can be used to limit forking and protect the original authors of the software. Developers ought to be careful in their choice of licence and consider institution or company policies, business models and the need to use copyleft libraries or link with proprietary software.

The use of software in education and training has three main learning objectives: learning to use a modelling software, investigating physical phenomena using modelling software and learning modelling algorithms and acquiring modelling skills. Learning to use modelling software is associated with the benefits of exploring hypothetical questions and scenarios without experimenting on the system itself. CSS can provide advantages, such as being stable and having well developed documentation, but it is often costly. However, OSS can motivate students and offer more transferable skills. Investigating physical phenomena using a modelling software can engage learners in active exploration and learning, sustain learning approaches and create situations that would be unacceptable in real life. CSS's strengths lie in high quality documentation and implemented numerical methods, which often guarantee convergence. OSS provides unique capabilities and facilitates the sharing and reuse of solvers, routines and exercises. Learning modelling algorithms and modelling skills is a more complex endeavour that requires implementing techniques in practical situations. OSS provides unique capabilities, but the steep learning curve can be a barrier in university curricula, making it more suitable for intensive and highly specialized courses.

It is argued that perhaps a 'default open' stance rather than a 'default closed' one would bring benefits to many scientific programmes worldwide; that decision, however, would need to be taken at the national level.

REFERENCES

[1] OPEN SOURCE INITIATIVE, The Open Source Definition (2024),
https://opensource.org/osd

[2] WILKINSON, M., et al., The FAIR Guiding Principles for scientific data management and stewardship, Sci. Data **3** (2016) 160018,
https://doi.org/10.1038/sdata.2016.18

[3] STALLMAN, R., Free Software, Free Society: Selected Essays of Richard M. Stallman, CreateSpace Independent Pub (2009).

[4] TORVALDS, L., DIAMOND, D., Just for Fun: The Story of an Accidental Revolutionary, HarperCollins, New York (2001).

[5] HAFNER, K., LYON, M., Where Wizards Stay Up Late: The Origins of the Internet, Simon & Schuster, New York (1999).

[6] CONTRERAS, J.L., JACOB, M. (Eds), Patent Pledges: Global Perspectives on Patent Law's Private Ordering Frontier, Edward Elgar Publishing Limited, Cheltenham (2017),
https://doi.org/10.4337/9781785362491

[7] MALSHE, D., Patent Pools, Competition Law and Biotechnology, Routledge, New York (2020).

[8] KNIGHT, H.J., Patent Strategy: For Researchers and Research Managers, 3rd edn, Wiley, Hoboken, NJ (2013),
https://doi.org/10.1002/9781118314289

[9] BAUWENS, M., KOSTAKIS, V., PAZAITIS, A., Peer to Peer: The Commons Manifesto, University of Westminster Press, London (2019),
https://doi.org/10.16997/book33

[10] RHODES, J., SMITH, K.S., LEE, D., "CASMO-5 Development and Applications", PHYSOR 2006, American Nuclear Society, Vancouver (2006).

[11] LOUBIÈRE, S., et al., "APOLLO2 twelve years later", International Conference on Mathematics and Computation, Reactor Physics and Environmental Analysis in Nuclear Applications (Proc. Int. Conf. Madrid, 1999), Senda Editorial, Madrid (1999).

[12] HÉBERT, A., DRAGON5 and DONJON5, the contribution of École Polytechnique de Montréal to the SALOME platform, Ann. Nucl. Energy **87** (2016) 12–20,
https://doi.org/10.1016/j.anucene.2015.02.033

[13] HERMAN, M., TRKOV, A. (Eds), ENDF-6 Formats Manual: Data Formats and Procedures for the Evaluated Nuclear Data Files ENDF/B-VI and ENDF/B-VII, Report BNL-90365-2009 Rev. 1, Brookhaven National Laboratory, Upton, NY (2010),
https://www.oecd-nea.org/dbdata/data/manual-endf/endf102.pdf

[14] COOPER, J.R., DOOLEY, R.B., The International Association for the Properties of Water and Steam, Report IAPWS R7-97, International Association for the Properties of Water and Steam, Lucerne (2012).

[15] GITHUB, A Curated List of Open Source Projects Used in Nuclear Science and Engineering (2023),
https://github.com/paulromano/awesome-nuclear

[16] TOURAN, N.W., et al., Computational tools for the integrated design of advanced nuclear reactors, Eng. **3** 4 (2017) 518–526,
https://doi.org/10.1016/J.ENG.2017.04.016

[17] BROOKS, H., DIXON, S., DAVIS, A., "Towards multiphysics simulations of fusion breeder blankets", International Conference on Physics of Reactors 2022 (PHYSOR 2022) (Proc. Int. Conf. Pittsburgh, 2022), American Nuclear Society, La Grange Park, IL (2022),
https://doi.org/10.13182/PHYSOR22-37726

[18] NOVAK, A.J., et al., "Coupled Monte Carlo and thermal-hydraulics modeling of a prismatic gas reactor fuel assembly using cardinal", International Conference on Physics of Reactors 2022 (PHYSOR 2022) (Proc. Int. Conf. Pittsburgh, 2022), American Nuclear Society, La Grange Park, IL (2022).

[19] MERZARI, E., et al., Cardinal: A lower-length-scale multiphysics simulator for pebble-bed reactors, Nucl. Technol. **207** 7 (2021) 1118–1141,
https://doi.org/10.1080/00295450.2020.1824471

[20] KELM, S., et al., The tailored CFD package 'containmentFOAM' for analysis of containment atmosphere mixing, H_2/CO mitigation and aerosol transport, Fluids **6** 3 (2021) 100,
https://doi.org/10.3390/fluids6030100

[21] LIU, X., et al., Monte Carlo method with SNBCK nongray gas model for thermal radiation in containment flows, Nucl. Eng. Des. **390** (2022) 111689, https://doi.org/10.1016/j.nucengdes.2022.111689

[22] FIORINA, C., CLIFFORD, I., KELM, S., LORENZI, S., On the development of multi-physics tools for nuclear reactor analysis based on OpenFOAM®: State of the art, lessons learned and perspectives, Nucl. Eng. Des. **387** (2022), https://doi.org/10.1016/j.nucengdes.2021.111604

[23] HUFF, K.D., et al., Fundamental concepts in the Cyclus nuclear fuel cycle simulation framework, Adv. Eng. Softw. **94** (2016) 46–59, https://doi.org/10.1016/j.advengsoft.2016.01.014

[24] DAGMC: Direct Accelerated Geometry Monte Carlo Toolkit (2023), http://svalinn.github.io/DAGMC/

[25] TAUTGES, T.J., WILSON, P.P.H., KRAFTCHECK, J., SMITH, B.M., HENDERSON, D.L., "Acceleration techniques for direct use of CAD-based geometries in Monte Carlo radiation transport", International Conference on Mathematics, Computational Methods & Reactor Physics 2009 (M&C 2009) (Proc. Int. Conf. Saratoga Springs, 2009), American Nuclear Society, Red Hook, NY (2009).

[26] ETTNER, F., VOLLMER, K.G., SATTELMAYER, T., Numerical simulation of the deflagration-to-detonation transition in inhomogeneous mixtures, J. Combust. **2014** (2014) 686347, https://doi.org/10.1155/2014/686347

[27] RITTER, C., et al., Versatile test reactor open digital engineering ecosystem, INSIGHT **25** (2022) 56–60, https://doi.org/10.1002/inst.12374

[28] RITTER, C., et al., Digital twin to detect nuclear proliferation: A case study, J. Energy Resour. Technol. **144** 10 (2022) 102108, https://doi.org/10.1115/1.4053979

[29] BROWNING, J., et al., Foundations for a fission battery digital twin, Nucl. Technol. **208** 7 (2022) 1089–1101, https://doi.org/10.1080/00295450.2021.2011574

[30] HEBERT, A., Applied Reactor Physics, 3rd edn, Les Presses de l'Université de Montréal, Montreal (2020), https://doi.org/10.1515/9782553017445

[31] MARLEAU, G., HEBERT, A., ROY, R., A User Guide for Dragon Version 5, Report IGE-335, Ecole Polytechnique Montréal, Montreal (2022).

[32] ROMANO, P., et al., A code-agnostic driver application for coupled neutronics and thermal-hydraulic simulations, Nucl. Sci. Eng. **195** (2020), https://doi.org/10.1080/00295639.2020.1830620

[33] POVILAITIS, M., JASELIŪNAITĖ, J., flameFoam: An open-source CFD solver for turbulent premixed combustion, Nucl. Eng. Des. **383** (2021) 111361, https://doi.org/10.1016/j.nucengdes.2021.111361

[34] JASELIŪNAITĖ, J., POVILAITIS, M., STUČINSKAITĖ, I., RANS-and TFC-based simulation of turbulent combustion in a small-scale venting chamber, Energies **14** (2021) 5710, https://doi.org/10.3390/en14185710

[35] POVILAITIS, M., JASELIŪNAITĖ, J., Simulation of hydrogen-air-diluents mixture combustion in an acceleration tube with FlameFoam solver, Energies, **14** 17 (2021) 5504,
https://doi.org/10.3390/en14175504

[36] PISSO, I., et al., The Lagrangian particle dispersion model FLEXPART version 10.4, Geoscientific Model Dev. **12** 12 (2019) 4955–4997,
https://doi.org/10.5194/gmd-12-4955-2019

[37] STOHL, A., et al., The Lagrangian particle dispersion model FLEXPART version 6.2, Atmos. Chem. Phys. **5** 9 (2005) 2461–2474,
https://doi.org/10.5194/acp-5-2461-2005

[38] STOHL, A., HITTENBERGER, M., WOTAWA, G., Validation of the Lagrangian particle dispersion model FLEXPART against large-scale tracer experiment data, Atmos. Environ. **32** 24 (1998) 4245–4264,
https://doi.org/10.1016/S1352-2310(98)00184-8

[39] ARNOLD, D., et al., Influence of the meteorological input on the atmospheric transport modelling with FLEXPART of radionuclides from the Fukushima Daiichi nuclear accident, J. Environ. Radioact. **139** (2015) 212–225,
https://doi.org/10.1016/j.jenvrad.2014.02.013

[40] FIORINA, C., CLIFFORD, I., AUFIERO, M., MIKITYUK, K., Gen-Foam: A novel Openfoam® based multi-physics solver for 2D/3D transient analysis of nuclear reactors, Nucl. Eng. Des. **294** (2015) 24–37,
https://doi.org/10.1016/j.nucengdes.2015.05.035

[41] FIORINA, C., KERKAR, N., MIKITYUK, K., RUBIOLO, P., PAUTZ, A., Development and verification of the neutron diffusion solver for the gen-foam multi-physics platform, Ann. Nucl. Energy **96** (2016) 212–222,
https://doi.org/10.1016/j.anucene.2016.05.023

[42] FIORINA, C., HURSIN, M., PAUTZ, A., Extension of the gen-foam neutronic solver to SP$_3$ analysis and application to the crocus experimental reactor, Ann. Nucl. Energy **101** (2017) 419–428,
https://doi.org/10.1016/j.anucene.2016.11.042

[43] RADMAN, S., FIORINA, C., MIKITYUK, K., PAUTZ, A., A coarse-mesh methodology for modelling of single-phase thermal-hydraulics of ESFR innovative assembly design, Nucl. Eng. Des. **355** (2019),
https://doi.org/10.1016/j.nucengdes.2019.110291

[44] RADMAN, S., FIORINA, C., PAUTZ. A., Development of a novel two-phase flow solver for nuclear reactor analysis: Algorithms, verification, and implementation in OpenFOAM, Nucl. Eng. Des. **379** (2021),
https://doi.org/10.1016/j.nucengdes.2021.111178

[45] RADMAN, S., FIORINA, C., PAUTZ, A., Development of a novel two-phase flow solver for nuclear reactor analysis: Validation against sodium boiling experiments, Nucl. Eng. Des. **384** (2021),
https://doi.org/10.1016/j.nucengdes.2021.111422

[46] IMRON, M., Development and verification of open reactor simulator ADPRES, Ann. Nucl. Energy **133** (2019) 580–588,
https://doi.org/10.1016/j.anucene.2019.06.049

[47] IMRON, M., HARTANTO, D., Pressurized water reactor mixed oxide/UO$_2$ transient benchmark calculations using Monte Carlo Serpent 2 code and open nodal core simulator ADPRES, ASME J. Nucl. Eng. Radiat. Sci. **7** 3 (2021) 031603,
https://doi.org/10.1115/1.4048764

[48] FRITZSON, P., et al., The OpenModelica integrated environment for modeling, simulation, and model-based development, Model. Identif. Control **41** 4 (2020) 241–285,
https://doi.org/10.4173/mic.2020.4.1

[49] PERMANN, C.J., et al., MOOSE: Enabling massively parallel multiphysics simulation, SoftwareX **11** (2020) 031603,
https://doi.org/10.1016/j.softx.2020.100430

[50] GASTON, D., NEWMAN, CH., HANSEN, G., LEBRUN-GRANDIE, D., MOOSE: A parallel computational framework for coupled systems of nonlinear equations, Nucl. Eng. Des. **239** 10 (2009) 1768–1778,
https://doi.org/10.1016/j.nucengdes.2009.05.021

[51] FISCHER, P., et al., NekRS, a GPU-accelerated spectral element Navier–Stokes solver, Parallel Comput. **114** (2022) 102982,
https://doi.org/10.1016/j.parco.2022.102982

[52] MacFARLANE, R.E., et al., The NJOY Nuclear Data Processing System, Version 2016, LANL report LA-UR-17-20093, Los Alamos National Security LLC, Los Alamos, TX (2017),
https://doi.org/10.2172/1338791

[53] STEWART, W.R., SHIRVAN, K., Capital cost estimation for advanced nuclear power plants, Renew. Sustain. Energy Rev. **155** (2021) 111880,
https://doi.org/10.1016/j.rser.2021.111880

[54] RADAIDEH, M.I., et al., NEORL: NeuroEvolution optimization with reinforcement learning — Applications to carbon-free energy systems, Nucl. Eng. Des. **412** (2023) 112423,
https://doi.org/10.1016/j.nucengdes.2023.112423

[55] RADAIDEH, M.I., et al., Physics-informed reinforcement learning optimization of nuclear assembly design, Nucl. Eng. Des. **372** (2021) 119066,
https://doi.org/10.1016/j.nucengdes.2020.110966

[56] RADAIDEH, M.I., FORGET, B., SHIRVAN, K., Large-scale design optimisation of boiling water reactor bundles with neuroevolution, Ann. Nucl. Energy **160** (2021) 108355,
https://doi.org/10.1016/j.anucene.2021.108355

[57] SCOLARO, A., CLIFFORD, I., FIORINA, C., PAUTZ, A., The OFFBEAT multi-dimensional fuel behavior solver, Nucl. Eng. Des. **358** (2020) 110416,
https://doi.org/10.1016/j.nucengdes.2019.110416

[58] CLIFFORD, I., et al., Studies on the effects of local power peaking on heat transfer under dryout conditions in BWRs, Ann. Nucl. Energy **130** (2019) 440–451,
https://doi.org/10.1016/j.anucene.2019.03.017

[59] SCOLARO, A., CLIFFORD, I., FIORINA, C., PAUTZ, A., "First steps towards the development of a 3D nuclear fuel behavior solver with OpenFOAM", Proc. of the 2018 26th Int. Conf. on Nuclear Engineering, ICONE26, London, 2018, ASME, London (2018),
https://doi.org/10.1115/ICONE26-82381

[60] SCOLARO, A., ROBERT, Y., FIORINA, C., CLIFFORD, I., PAUTZ A., "Coupling methodology for the multidimensional fuel performance code offbeat and the Monte Carlo neutron transport code SERPENT", Proc. of 14th Int. Nuclear Fuel Cycle Conf., GLOBAL 2019 and Light Water Reactor Fuel Performance Conf. TOP FUEL 2019, Seattle, 2019, ASME, La Grande Park, IL (2020).

[61] SCOLARO, A., et al., Investigation on the effect of eccentricity for fuel disc irradiation tests, Nucl. Eng. Technol. **53** 5 (2021) 1602–1611, https://doi.org/10.1016/j.net.2020.11.003

[62] DE TROULLIOUD DE LANVERSIN, J., KÜTT, M., GLASER, A., ONIX: An open-source depletion code, Ann. Nucl. Energy **151** (2021) 107903, https://doi.org/10.1016/j.anucene.2020.107903

[63] WELLER, H.G., TABOR, G., JASAK, H., FUREBY, C., A tensorial approach to computational continuum mechanics using object-oriented techniques, Comput. Phys. **12** 6 (1998) 620–631, https://doi.org/10.1063/1.168744

[64] GREENSHIELDS, C.J., WELLER H.G., Notes on Computational Fluid Dynamics: General Principles, CFD Direct Ltd, Reading (2022).

[65] ROMANO, P.K., et al., OpenMC: A state-of-the-art Monte Carlo code for research and development, Ann. Nucl. Energy **82** (2015) 90–97, https://doi.org/10.1016/j.anucene.2014.07.048

[66] ROMANO, P.K., FORGET, B., The OpenMC Monte Carlo particle transport code, Ann. Nucl. Energy **51** (2013) 274–281, https://doi.org/10.1016/j.anucene.2012.06.040

[67] OPENMC, Publications overview for OpenMC (2023), https://docs.openmc.org/en/latest/publications.html

[68] OPENMOC, Publications overview for OpenMOC (2023), https://mit-crpg.github.io/OpenMOC/publications.html

[69] BIONDO, E., et al., Quality assurance within the PyNE open source toolkit, Trans. Am. Nucl. Soc. **111** (2014) 1169–1172.

[70] BATES, C.R., et al., PyNE progress report, Trans. Am. Nucl. Soc. **111** (2014) 1165–1168.

[71] ALFONSI, A., et al., RAVEN Theory Manual, INL/EXT-16-38178, Idaho National Laboratory, Idaho Falls, ID (2020), https://doi.org/10.2172/1784873

[72] ALFONSI, A., et al., RAVEN User Guide, INL/EXT-18-44465, Idaho National Laboratory, Idaho Falls, ID (2018), https://doi.org/10.2172/1467401

[73] RABITI, C., et al., RAVEN User Manual, INL/EXT-15-34123, Idaho National Laboratory, Idaho Falls, ID (2017), https://doi.org/10.2172/1364096

[74] PIZZOCRI, D., BARANI, T., LUZZI, L., SCIANTIX: A new open-source multi-scale code for fission gas behaviour modelling designed for nuclear fuel performance codes, J. Nucl. Mater. **532** (2020) 152042, https://doi.org/10.1016/j.jnucmat.2020.152042

[75] ZULLO, G., et al., Towards grain-scale modelling of the release of radioactive fission gas from oxide fuel, Part I: SCIANTIX, Nucl. Eng. Technol. **54** 2 (2022) 2771–2782, https://doi.org/10.1016/j.net.2022.02.011

[76] BARANI, T., et al., Modeling high burnup structure in oxide fuels for application to fuel performance codes, Part II: Porosity evolution, J. Nucl. Mater. **563** (2022) 153627, https://doi.org/10.1016/j.jnucmat.2022.153627

[77] MAGNI, A., et al., Application of the SCIANTIX fission gas behaviour module to the integral pin performance in sodium fast reactor irradiation conditions, Nucl. Eng. Technol. **54** 7 (2022) 2395–2407, https://doi.org/10.1016/j.net.2022.02.003

[78] KOWALSKI, M., COSGROVE, P., BROMAN, J., SHWAGERAUS, E., SCONE: A student-oriented modifiable Monte Carlo particle transport framework, J. Nucl. Eng. **2** 1 (2021) 57–64, https://doi.org/10.3390/jne2010006

[79] KOWALSKI, M., SHWAGERAUS, E., A hybrid continuous energy and multi-group Monte Carlo method, Ann. Nucl. Energy **140** (2020) 107277, https://doi.org/10.1016/j.anucene.2019.107277

[80] KOWALSKI, M., COSGROVE, P., Acceleration of surface tracking in Monte Carlo transport via distance caching, Ann. Nucl. Energy **152** (2021) 108002, https://doi.org/10.1016/j.anucene.2020.108002

[81] RAFFUZZI, V., MORGAN, L., SHWAGERAUS, E., Accelerating Monte Carlo neutron transport by approximating thermal cross sections with functional forms, Ann. Nucl. Energy **169** (2022) 108819, https://doi.org/10.1016/j.anucene.2021.108819

[82] HELFER, T., et al., Introducing the open-source mfront code generator: Application to mechanical behaviors and material knowledge management within the PLEIADES fuel element modelling platform, Comput. Math. Appl. **70** (2015) 994–1023, https://doi.org/10.1016/j.camwa.2015.06.027

[83] MFRONT (2023), https://thelfer.github.io/tfel/web/publications.html

[84] TRIOCFD, Le code TrioCFD (2023), https://triocfd.cea.fr/Pages/Presentation/TrioCFD_code.aspx

[85] TRUST (2023), https://github.com/cea-trust-platform/trust-code

[86] SAIKALI, É., et al., "Numerical modeling of a moderate hydrogen leakage in a typical two-vented fuel cell configuration", Int. Conf. on Hydrogen Safety, Edinburgh (2021).

[87] DEMAZIÈRE, C., CORE SIM: A multi-purpose neutronic tool for research and education, Ann. Nucl. Energy **38** 12 (2011) 2698–2718, https://doi.org/10.1016/j.anucene.2011.06.010

[88] HASSLBERGER, J., KATZY, P., BOECK, L.R., SATTELMAYER, T., Computational fluid dynamics simulation of deflagration-to-detonation transition in a full-scale Konvoi-type pressurized water reactor, J. Nucl. Eng. Radiat. Sci. **3** 4 (2017) 041014, https://doi.org/10.1115/1.4037094

[89] BARFUSS, C., HEILBRONN, D., SATTELMAYER, T., "Simulation of deflagration-to-detonation transition of lean H_2-CO-air mixtures in obstructed channels", Int. Conf. on Hydrogen Safety (2019).

[90] ZIVKOVIC, D., SATTELMAYER, T., "Towards efficient and time-accurate simulations of early stages of industrial scale explosions", Int. Conf. on Hydrogen Safety (2021).

[91] KASSELMANN, S., et al., Verification of the HTR code package (HCP) as a comprehensive HTR steady state and transient safety analysis framework, Nucl. Eng. Des. **329** (2018) 167–176,
https://doi.org/10.1016/j.nucengdes.2017.11.044

[92] TANTILLO, F., et al., HTR code package neutronics developments and benchmarks, Nucl. Eng. Des. **362** (2020) 110603,
https://doi.org/10.1016/j.nucengdes.2020.110603

[93] KASSELMANN, S., et al., Status of the development of a fully integrated code system for the simulation of high temperature reactor cores, Nucl. Eng. Des. **271** (2014) 341–347,
https://doi.org/10.1016/j.nucengdes.2013.11.059

[94] HELMHOLTZ ZENTRUM DRESDEN ROSSENDORF, Sustainable Development of Simulation Software for Modeling of Reactor Coolant Systems (2023),
https://www.hzdr.de/openfoam-rcs

[95] JEON, S., HONG, H., CHOI, N., JOO, H.G., "GPU acceleration of the prototype pinwise core analysis code VANGARD", Int. Conf. on Mathematics and Computational Methods Applied to Nuclear Science and Engineering (Proc. Int. Conf. Virtual Meeting, 2021), American Nuclear Society, La Grange Park, IL (2021).

[96] JEON, S., JOO, H.G., "Verification and validation of the GPU-based high-speed pinwise nodal core analysis code VANGARD", Transactions of the Korean Nuclear Society Spring Meeting, Jeju (2022).

[97] SWARTS, J., Open-source software in the sciences: The challenge of user support, J. Bus. Tech. Commun. **33** (2019) 60–90,
https://doi.org/10.1177/1050651918780202

[98] NumPy, Berkeley Institute for Data Science (2015),
https://bids.berkeley.edu/numpy

[99] HICKS, M., Built to last, Logic Magazine (31 Aug. 2020) 11.

[100] AMERICAN SOCIETY OF MECHANICAL ENGINEERS, Quality Assurance Requirements for Nuclear Facility Applications, American National Standards Institute, New York (2022).

[101] NUMFOCUS, Mission of NumFOCUS (2023),
https://numfocus.org/community/mission

[102] LINDSAY, A.D., et al., MOOSE: Enabling massively parallel multiphysics simulation, SoftwareX **20** (2022) 101202,
https://doi.org/10.1016/j.softx.2022.101202

[103] Communications Received from Certain Member States Regarding Guidelines for the Export of Nuclear Material, Equipment or Technology, INFCIRC/254 Part 1, IAEA, Vienna (1978).

[104] Communications Received from Certain Member States Regarding Guidelines for the Export of Nuclear Material, Equipment or Technology, INFCIRC/254 Part 2, IAEA, Vienna (1978).

[105] NATIONAL SECURITY DECISION DIRECTIVES, National Policy on the Transfer of Scientific, Technical and Engineering Information (1985),
https://irp.fas.org/offdocs/nsdd/nsdd-189.htm

[106] NATIONAL ARCHIVES, Code of Federal Regulations, Commerce and Foreign Trade, Title 15, Subtitle B, Chapter VII, Subchapter C, §734.8(c) (2001).

[107] NATIONAL ARCHIVES, Code of Federal Regulations, Energy, Title 10, Chapter III, §810.3 (2012).

[108] NUCLEAR ENERGY AGENCY, IRPhe Handbook 2021, International Reactor Physics Evaluation Project Handbook (2023),
https://www.oecd-nea.org/jcms/pl_20279/international-handbook-of-evaluated-reactor-physics-benchmark-experiments-irphe

[109] O'DELL, R.D., Standard Interface Files and Procedures for Reactor Physics Codes, Report 5369298, Version IV, Los Alamos, NM, Los Alamos National Laboratory (1977),
https://doi.org/10.2172/5369298

[110] WORLD INTELLECTUAL PROPERTY ORGANIZATION, Summary of the Berne Convention for the Protection of Literary and Artistic Works, WIPO (1886),
https://www.wipo.int/treaties/en/ip/berne/summary_berne.html

Annex

SUMMARY OF THE TECHNICAL MEETING

This Annex provides summaries of the different sessions and papers presented at the Technical Meeting on the Development and Application of Open Source Modelling and Simulation Tools for Nuclear Reactors, as well as summaries of discussion panels.[1] The technical meeting also provided the participants with an opportunity to take part in the workshops covering a range of open source tools, such as Advanced Reactor Modeling Interface (ARMI), OpenFOAM, OpenMC and MOOSE.

A–1. GENERAL SESSIONS 1–1, 1–2 AND 1–3: MOTIVATIONS, EXPERIENCE AND CHALLENGES FOR THE OPEN-SOURCE APPROACH

There are numerous motivations for the use of open source software (OSS) within the field of nuclear engineering. Since open source tools are free to use, distribute and modify, they may be adopted easily to teach new concepts to students of academic institutions. Furthermore, the exposure of the source code may be conducive to a deeper understanding of the underlying physics, numerical methods and algorithms employed in simulations. By training students in a given toolset, its later use in industry or research to solve real-world problems is facilitated; furthermore, it is easier to integrate open source tools and eventually create automated workflows. In contributing towards OSS projects, students also gain transferable skills. Finally, OSS development is supported by a large global community that nurtures newer members via forums and can accelerate discovery of novel solutions while avoiding potential duplication of work.

Notwithstanding the many benefits surrounding the use of OSS, there are also many challenges. In nuclear engineering, where safety is paramount, there are concerns over trustworthiness and usability. There is also a need to protect from malicious intent and ensure that certain intellectual property is protected. Nevertheless, lessons learned from many years of OSS development in nuclear engineering offer some solutions to these challenges.

---

[1] The papers are available on the dedicated Indico site: https://conferences.iaea.org/event/247/

Paper titles are reproduced from the original material as submitted by the contributors and have not been edited by the editorial staff of the IAEA.

It is possible to ensure QA by adhering to best practices in software development, such as the use of version control, testing, continuous integration and containerization, tracking of issues, tagging of releases with accompanying change logs, and provision of user and developer documentation. Modularity of code design and appropriate licence selection enable the isolation of sensitive information, such as data, validation schemes, benchmarks or designs.

Providing adequate training in the use of the software helps both industry and academia to engage and grow the number of skilled users. Different teaching formats promote different styles of learning. Live workshops and hackathons (both virtual and in person) can overcome initial technical barriers, teach patterns of problem solving and signpost other materials. This approach is complemented by tutorials, examples, on-line courses and forums that may be employed for self-guided learning. In both industry and academia, it is important that such materials remain up to date and consistent with the latest version of the code; where possible, this is ensured by automated generation of such material. Finally, creating a kind and inclusive culture will allow open conversation, exchange of ideas and identification of novel solutions to problems.

A–1.1. The GEN-Foam multi-physics solver as a collaborative effort towards an open-source platform for reactor analysis: a historical perspective and lessons learnt

Generalized Nuclear Foam (GeN-Foam) is an OSS based on OpenFOAM used for steady state and transient analysis of nuclear reactors. The development of the software started in 2014 at the Paul Scherrer Institute in Switzerland. GeN-Foam became a central development effort for the Laboratory for Reactor Physics and Systems Behaviour at the Swiss Federal Institute of Technology Lausanne in 2015. Several institutions have collaborated in the development of GeN-Foam and its capabilities, such as development of the discussion sub-solver, the point kinetics solver for molten salt reactors and others.

GeN-Foam has established a Git code versioning system to allow collaborative development with tracked modifications and various code branches. Object oriented programming has also been adopted, dividing the code into classes that isolate different functionalities. A private discussion forum was opened to facilitate communication among users and developers, and a regression test was introduced in 2018 to verify the consistency of results after code modifications. The availability of tutorials and more comprehensive documentation led to the first beta release of GeN-Foam in 2020, followed by an official public release in 2021. The adoption of object oriented programming has limited maintenance efforts, allowing modifications to be made without affecting other classes. However, specific parts of the code have broken encapsulation using a single mechanism.

A–1.2. Contributing to OpenFOAM Foundation release – an effective code development strategy

Open source software has had a significant impact on the field of software development over recent decades, with many examples of successful OSS such as the Linux kernel, LibreOffice and Python. In the specialized field of computational fluid dynamics (CFD), OpenFOAM has emerged as the leading OSS package for numerical simulations in engineering applications, including nuclear safety research. The transparency and reliability provided by access to the source code is a crucial aspect of verifying the accuracy of methods and models used in CFD simulations, ensuring long term availability and independence from commercial interests of software manufacturers. The use of OSS also enables transparency and reproducibility of research results, which is a fundamental requirement for good scientific work. However, the maintenance and further development of OSS requires a significant number of users and the adherence to criteria such as robustness, functionality, usability, extensibility and accessibility. OpenFOAM is designed as a library with multiple solvers, utilities and model libraries, which allows a wide range of applications, depending on the physical problem to be analysed.

OpenFOAM is an OSS package that provides a basis for the community to develop new technologies and extensions. However, many users divide a stable version of OpenFOAM for their implementations, which creates problems such as unsynchronized code, unrecognized bugs and difficulties in sharing code. Therefore, contributing to the OpenFOAM Foundation release is desirable, but it requires a high standard of software design, robustness, code style and maintainability. Investing resources in contributions and discussions with core maintainers helps in developing OpenFOAM in a sustainable and maintainable way. Helmholtz-Zentrum Dresden-Rossendorf and VTT Technical Research Centre of Finland are active contributors to the OpenFOAM Foundation release and have established maintenance plans and support contracts for their extensions to ensure effective and efficient development of new technologies in the field of CFD. Other contributors have also developed new functionality relevant for nuclear applications. Robustness, functionality, usability, extensibility and accessibility are important criteria for the sustainable development of a complex OSS package such as OpenFOAM.

A–1.3. DRAGON and DONJON: a legacy open-source reactor physics project

DRAGON is a lattice code that performs deterministic and multigroup solution of neutron and photon transport equations. It is compatible with the DONJON full-core simulation code, providing a range of capabilities including neutron diffusion, cross-section interpolation, depletion, critical control and thermal hydraulics. Each constitutes a set of modules that may be combined into computational schemes using the CLE-2000 scripting language. These tools have had significant industrial usage over the years, in particular for the modelling of the Canada deuterium–uranium (CANDU) reactor, a Canadian pressurized heavy water reactor, since 2006.

DRAGON, now in its fifth major version, has an extensive history of development spanning 40 years. It was made open source in 1989 under the GNU Lesser General Public License (LGPL), and was an early adopter of many QA practices, such as the use of version control software with the release of Apache Subversion (often known as SVN) in 2003. Other such practices include testing, documentation, issue tracking, configuration management and tagging of releases accompanied by a change log; these have played an important role in the long term maintenance and sustainability of the project.

In addition to QA, the adoption of DRAGON and DONJON by the nuclear industry has been facilitated by their modularization and specific choice of the open source LGPL. This licence applies only to the portion of the code to which it has already applied, not to the entirety of any derivative programs. Thus, any derived computational schemes may be still subject to intellectual property restrictions. However, requiring users to devise such schemes mandates that they have a deeper understanding of the code, which may simultaneously be viewed as both a challenge and a benefit.

A–1.4. On application of open-source learning models to nuclear engineering

A selection of lessons learned from teaching OSS to the nuclear engineering community was presented. Many of these insights have been taken from one organization in particular, The Carpentries[2], which provides software skill development material hosted on collaborative open source platforms in a model known as Software Carpentry. Through a culture of inclusivity and the cultivation of teaching material with methods similar to those employed in software development, upskilling is accelerated and there can be a short life cycle from learner to instructor.

The Software Carpentry model of teaching is one of two complementary approaches adopted by the OpenMC community, with the second approach being self-guided learning. In the first approach, instructor-led workshops are used to educate learners in how to teach themselves while overcoming initial

---

[2] The Carpentries (https://carpentries.org/) is a non-profit organization that teaches foundational coding and data science skills to researchers worldwide.

technical issues that might otherwise block progress. An additional benefit is the requirement for a learner to block out a dedicated slot of time to attend. By contrast, self-guided material is always available and can be very effective when combined with a community forum.

Many aspects need to be considered to ensure that workshops are successful. The environment ought to be free from distractions; gathering information before, during and after the session helps to gauge the progress of attendees; cognitive load needs to be managed appropriately; and consistent and stable installations of software — for example, with containers — need to be created. With self-guided training, it is important that the material is sufficiently broken down into easily digestible modules and that it remains up to date with the latest version of the code. Keeping the material up to date can be challenging where there is rapid development; therefore, it is desirable that this be automated where possible.

A–1.5. The modern Nuclear Energy Agency Data Bank: An integrated hub for code and data development and validation

The Nuclear Energy Agency (NEA) Data Bank is a centre of reference for codes, nuclear data, benchmarks, training and knowledge preservation, and has three core missions. The Computer Program Service (CPS) handles acquisition, licensing, testing and distribution of computer codes, and organization of training courses. The Nuclear Data Service compiles measured nuclear reaction data (EXFOR library) and coordinates the Joint Evaluated Fission and Fusion (JEFF) project. Finally, the NEA Data Bank preserves and distributes experimental data and provides support for training and educational activities.

The NEA Data Bank has recently made improvements in its platforms and has modernized many of its practices, with the intention of integrating its services and ultimately automating the workflow to validate codes against benchmarks for given nuclear data. The objective is to move towards ways of working that focus on transparency, collaboration and reproducibility. To fulfil this objective, technical engagement is necessary; this is being enacted through a series of technical focus groups culminating in a hackathon.

Along with the overhaul in its hosting and pipeline of codes and data, the NEA Data Bank is now providing e-learning services to support education and training in the use of the tools that it hosts. These modularized courses have both a virtual taught component and on-line content containing exercises and assessments.

A–1.6. Use of open-source tools in education and training: Experience from POLIMI and EPFL

Bloom's taxonomy of the cognitive domain classifies learning objectives into six hierarchical categories, by order of increasing complexity: knowledge, comprehension, application, analysis, evaluation and creation. According to this categorization, higher order cognitive skills build upon the foundational lower order ones, the latter being required in greater volume. Recognizing this distinction, it is important to select a teaching format appropriate for the desired outcome of the training.

While foundational knowledge might be the traditional remit of the classroom, numerous OSS tools may be effectively utilized to link concepts and recursively combine different competencies and guide a student through comprehension to analytical and creative approaches to problem solving. Examples from the Polytechnic University of Milan (PoliMi) and the Swiss Federal Institute of Technology Lausanne (EPFL) include the use of the following:

— OpenFOAM to teach the setup and solution of numerical simulations;
— Jupyter notebooks to teach the numerical equations relating to reactor physics;
— Modelica to teach a systems approach to nuclear engineering of reactors.

A–1.7. Challenges and lessons learned from 10 years of OpenMC development

This presentation provided reflections on the OpenMC code, which has been under continuous development for over ten years and now boasts a thriving community of users and developers. As OpenMC is the first attempt at a community developed Monte Carlo code for nuclear energy applications, the development team has faced many challenges, including funding maintenance and user support, the lack of organizational ownership, the time required for unsupported work (e.g. communication, user requests), the lack of traditional success metrics, talent acquisition, technical debt, handling backward incompatible code changes and maintaining performance. Despite these challenges, the open source model has been a key part of OpenMC's success and longevity by enabling wider engagement of users and developers who have helped to steer the code development in a positive direction. One of the key lessons learned by the OpenMC team is that while it is easy to make a project open source, it takes years to build enough momentum to have a truly sustainable community.

A–1.8. Use of open source software at ENEA for nuclear reactor safety simulations

The application oriented presentation introduced cases of OSS applied to nuclear reactor safety issues at the Italian National Agency for New Technologies, Energy and Sustainable Economic Development (ENEA). Different OSS packages were customized and applied for specific applications. The OpenFOAM CFD package was extended by a turbulent heat transfer model to analyse liquid metal flows in liquid metal cooled fast reactors and molten corium. The libMesh finite element method (FEM) package was employed to implement an arbitrary Lagrangian–Eulerian formulation to analyse fluid–structure interactions during water injection into, and dust remobilization during loss of vacuum accidents in ITER was simulated using a Euler–Lagrangian approach using the code_Saturne. The dispersion of (radioactive) pollutants on the mesoscale was simulated using the FLEXPART code. A set of OSS helper tools (FLEXTools) was developed to aid its use. Coupled effect analysis was performed using the SALOME platform and the ICoCo interface. Furthermore, the MEDUSA platform was developed on the basis of SALOME's MED and MEDCoupling libraries to couple codes and enable multiphysics and multiscale analysis involving neutronics, thermomechanics and thermal hydraulics. Several OSS tools, such as Gmsh and ParaView, were used for pre and postprocessing. The individual extensions and developments have been partially published and all of them will become available in the future.

In concluding, the authors stated that OSS enables easier and powerful collaboration with universities within degree related research. The advantages are engagement of the students in the OSS communities, the possibility to share or publish the results, the option to identify small self-contained tasks for students and the portability of the obtained skills and methods. In the context of nuclear safety, the development of open source system level codes as well as two-phase flow models for relevant flow regimes will be particularly required in the future. The use of OSS enables compliance with the findable, accessible, interoperable and reusable (FAIR) and open science principles and thus facilitates the acquisition of public project funds (e.g. Euratom or Horizon Europe).

The discussion addressed the issue of export control. It was concluded that there is no issue related to the distribution of fundamental libraries and original packages; however, any proprietary application or design specific correlations (e.g. pressure drop, heat transfer) are not to be published.

A–1.9. SCONE: An open source Monte Carlo neutron transport code for research and teaching

The author introduced a new teaching and educational toolbox, known as SCONE, and shared the reasons for its development and the latest features, and presented a number of success stories where students had used the tool for their research at the undergraduate, Masters and PhD levels.

The discussion involved the following points:

— Involvement of students. If more students work at the same time, they can support each other to conclude their assignments successfully.
— The reason for the open source model and associated legal issues.
— Whether the various features of SCONE will be maintained or archived if not further used.
— The timeliness of governance models.

A–2. TECHNICAL SESSIONS 1–1 AND 1–2: ADVANCED APPLICATION OF OPEN-SOURCE TOOLS TO NUCLEAR ENGINEERING PROBLEMS

A–2.1. A framework for multi-physics modeling, design optimization and uncertainty quantification of fast spectrum liquid fuel molten salt reactors

Most modelling and simulation tools for nuclear reactors have been focused on light water reactors owing to their extensive operational experience. However, there has been recent development of tools for modelling Generation III, III+, IV type systems, including liquid fuel molten salt reactors (LFMSRs). LFMSRs present unique multiphysics challenges with strong coupling between thermal hydraulics, neutronics, inventory control and species distribution phenomena. In order to advance the development of LFMSR concepts, a framework for multiphysics modelling, design optimization and uncertainty quantification is needed. The GeN-Foam and System Analysis Module (SAM) codes were used in this work to analyse an LFMSR, with an hourglass shaped core geometry chosen to avoid recirculation zones. CFD and system level modelling are necessary to explore the impact of phenomena on the reactor system during steady state and transient scenarios. Quantifying uncertainties related to design parameters is crucial in detailed studies of the reactor geometry. The study also aimed to define a safe operating zone for the LFMSR design by accounting for the limiting safety system settings. The LFMSR design comprised 16 loops and the thermophysical properties used to set up the case were listed. Future work will focus on testing the reactor design under transient scenarios and performing high fidelity multiphysics 2-D primary loop uncertainty and sensitivity analysis using the GeN-Foam–Dakota coupling.

A–2.2. Use of open-source Modelica-based libraries as system code for nuclear power plant simulation

Thermal hydraulic system codes play a crucial role in the analysis of nuclear systems, especially for evaluating operational and accident scenarios. However, because of the considerable resources involved in validating these tools and developing the necessary correlations, such codes are typically proprietary and subject to licence restrictions. In the long term, open source alternatives are worth pursuing, particularly for modelling ex-core components. The Modelica language is a modern modelling language for the simulation of engineering and physical systems that is well suited for use as a system thermal hydraulics code in the nuclear field. The advantages of Modelica include its object oriented and declarative nature, equation based approach, acausal modelling and component oriented structure.

Using Modelica enables the creation of libraries of validated power plant components that can be reused across different plant configurations. Modelica based modelling and the use of already validated components can facilitate the development of ad hoc correlations suitable for nuclear plant conditions, enabling an open data paradigm for verification and validation (V&V) processes, and facilitating error detection. Other advantages include the potential for coupling with other modelling approaches via the Functional Mock-up Interface. Open source Modelica based libraries have been developed for use in a range of fields, including nuclear simulation. Examples include the ThermoPower library for the dynamic modelling of thermal power plants and energy conversion systems, the NuKomp library of nuclear components for the modelling of nuclear power plants, and other libraries developed at the Polytechnic University of Milan for power plant simulator control purposes, natural circulation phenomena and more.

In the short term, it is not feasible to substitute deterministic safety analysis in existing reactors with open source alternatives. However, in the longer term, developing open source alternatives to legacy codes is essential for the nuclear industry to address the challenge of validating existing models and enhancing flexibility in model development. By using Modelica and other modern technologies, nuclear safety analysts can benefit from a more open and flexible approach to simulation, which can be tailored to specific applications and adapted to evolving technologies. Additionally, open source codes can facilitate collaboration and the sharing of knowledge within the thermal hydraulics community.

A–2.3. flameFOAM: An OpenFOAM based solver for practical turbulent premixed combustion simulation

Nuclear power plant containment is the last line of defence against a radioactive release during a severe accident, and hydrogen combustion poses the highest risk to containment integrity. As such, there is an international effort to predict hydrogen combustion phenomena in nuclear power plant containments during severe accidents. Modelling hydrogen combustion is challenging, particularly simulating flame acceleration due to turbulence. Most simulations in the scientific literature and international benchmarks are performed using commercial codes with in-house modified combustion models or proprietary solutions, with open source solutions representing only a small fraction. This domination of proprietary solutions stifles innovation, promotes fragmentation and creates challenges for newcomers. A freely accessible open source solver would promote knowledge dissemination and lower barriers to modelling turbulent premixed combustion.

To address these issues in containment analysis, the authors of a study selected the OpenFOAM CFD toolkit to create an open source turbulent premixed combustion solver called flameFoam. OpenFOAM is widely used for the simulation of compressible and incompressible fluid mechanics, heat and mass transfer, and reacting flows. flameFoam is based on standard OpenFOAM solvers and uses the progress variable approach with the turbulent flame closure model expression for the combustion source term. The laminar flame speed value can be given by the user or can be dynamically calculated using Malet's correlation. flameFoam can distinguish between fluid and solid regions in a simulation domain, and conjugate heat transfer between them is realized using standard OpenFOAM conditions. The present version of flameFoam has been developed and tested for the combustion of homogeneous hydrogen–air mixtures, and the model features implemented are equivalent to the state of the art demonstrated in recent international benchmarks.

A–2.4. Open-sourcing safety, risk, and reliability tools

The authors presented the development of several software tools (RAVEN, LOGOS and the Safety Risk Reliability Model Library (SR2ML)) designed to perform safety, risk and reliability analysis of power plants that are currently under development and being used for RISA applications.

RAVEN is a software framework that is designed to perform parametric and stochastic analyses based on the response of complex systems codes. It can communicate directly with the system codes described above, which are currently used to perform plant safety analyses. The provided application programming interfaces allow RAVEN to interact with any code if all the parameters that need to be perturbed are accessible by input files or via Python interfaces. RAVEN is capable of investigating system response and exploring input spaces using various sampling schemes, such as Monte Carlo, grid or Latin hypercube. However, RAVEN's strength lies in its system feature discovery capabilities, such as constructing limit surfaces, separating regions of the input space leading to system failure and using dynamic supervised learning techniques.

SR2ML is a software package that contains a set of reliability models designed to be interfaced with the RAVEN code, developed at the Idaho National Laboratory. These models can be employed to perform both static and dynamic system risk analysis and determine risk importance of specific elements of the considered system. Two classes of reliability models have been developed. The first class includes all classical reliability models (i.e. fault trees, event trees, Markov models and reliability block diagrams), which have been extended to deal not only with Boolean logic values but also time dependent values. The second class includes several component ageing models. Models included in these two classes are designed to be included in a RAVEN ensemble model to perform time dependent system reliability analysis (e.g. dynamic analysis). Similarly, these models can be interfaced with system analysis codes within RAVEN to determine the failure time of systems and evaluate accident progression (static analysis).

LOGOS contains a set of discrete optimization models that can be employed for capital budgeting optimization problems. LOGOS integrates economic and reliability risk in a single analysis framework. More specifically, provided structure, system and component (SSC) health information (e.g. failure rate, failure probability), operational and maintenance costs, replacement costs, cost associated with component failure and plant budget constraints, LOGOS provides the optimal set of projects (e.g. SSC replacement or refurbishment) that maximizes profit and satisfies the provided requirements. The input data listed above can be either deterministic or stochastic in nature, that is, they can be point values or probability distribution functions. In the latter case, several scenarios are generated by sampling of the provided distributions. The developed models are based on different versions of the knapsack optimization problem. Two main classes of optimization models were initially developed: deterministic and stochastic. Stochastic optimization models advance deterministic models by explicitly considering data uncertainties (associated to constraints or item cost and reward). These models can be employed as standalone models or can be interfaced with the RAVEN code to propagate data uncertainties and analyse the generated data (i.e. sensitivity analysis).

These projects have been developed using GitHub or GitLab as a software development and version control platform, which allows developers to track and support not only source code development but also code verification. In particular, code verification is performed every time the source code is updated or changed through series of regression and unit tests. Code requirements are directly linked to a subset of regression and unit tests. Documentation development (e.g. user manuals, user guides, workshop material) follows the same standards (e.g. documentation compilation) and is updated to reflect changes in the source code.

A–2.5. MFEM-MGIS-MFRONT, a HPC mini-application targeting nonlinear thermo-mechanical simulations of nuclear fuels at mesoscale

The context of this work is twofold: (a) valorization of the open source SALOME platform for performing thermal hydraulics studies and (b) integration of the mini application with the SALOME platform and support of the MED (modèle d'échange de données) mesh file format.

The MFEM-MGIS-MFRONT mini application aims at efficiently using supercomputers in the field of implicit non-linear thermomechanics. This open source library is based on several components as prerequisites. The first component, MFEM, is a FEM library designed for current supercomputers but also for the upcoming exascale supercomputers. It provides many useful features for carrying out realistic simulations: support for curvilinear meshes, high order approximation spaces and different families of finite elements, interfaces to several types of parallel solver (including matrix-free ones), preconditioners and native support for adaptive non-conforming mesh refinement.

Originating from the applied mathematics and parallel computing communities, MFEM offers both performance and a large panel of advanced mathematical features. In particular, it is possible to easily switch from one linear solver to another (direct or iterative), which is essential for a targeted application: microstructure and mesoscale modelling for nuclear fuel. However, applications to solid mechanics in MFEM are mostly limited to simple constitutive equations such as elasticity and hyperelasticity, which is insufficient.

A–3. TECHNICAL SESSIONS 2–1 AND 2–2: RECENT DEVELOPMENTS IN OPEN-SOURCE TOOLS

A–3.1. Updates on open-source multiphysics endeavors with OpenMC

The author provided insights into the development, performance and future activities of the code. OpenMC is a community driven Monte Carlo code that originated at the Massachusetts Institute of Technology and is now being led by

Argonne National Laboratory. The code is utilized for reactor physics simulations and includes modules for reactor depletion simulations. The inclusion of the depletion module enhances the use of the code for reactor design and analyses. A key feature of OpenMC is the ability to utilize the code on high performance computing platforms to solve high fidelity and highly complex systems encountered in fission and fusion devices. The code can use central processing unit (CPU) or graphic processing unit (GPU) based systems. This capability enhancement has been supported by the Exascale Computing programme of the United States Department of Energy (USDOE). The author presented an overview of two projects that are supporting the enhancements in and use of OpenMC: the Exascale Nuclear Reactor Investigative Code (ENRICO) and A Unified Resource for OpenMC (fusion) Reactor Applications (AURORA). Challenges in developing multiphysics codes were also presented along with recommendations to address these issues.

A–3.2. High performance calculation enhancement using massive parallelization with TRUST and TrioCFD

The author provided a detailed update on the development of TRUST and TrioCFD. TRUST is an open source framework for fluid mechanics and thermal hydraulics development, whereas TrioCFD is a CFD code based on the TRUST platform. The TRUST platform provides for the integration of multiple computational models for system thermal hydraulics, core thermal hydraulics and chemical reactions. The TrioCFD module of TRUST performs solutions of the Navier–Stokes equations with increased complexity including turbulent flow models, fluid–structure interactions, multiphase flows and flows in porous media. Enhancements to TRUST and TrioCFD were incorporated to improve performance on high performance computing platforms, allowing billions of cells to be modelled in the simulations as well as the implementation of 64-bit coding. Memory optimization has also been improved to enable simulations with billions of cells and tens of thousands of processors.

A–3.3. Development of the open-source multi-dimensional fuel performance code OFFBEAT

The author presented work on the development of the OpenFOAM Fuel BEhavior Analysis Tool (OFFBEAT). OFFBEAT was developed as a research tool to better model the physical behaviour of reactor fuels, including the presence of non-uniform oxide, crud formation and pellet–cladding mechanical interaction. OFFBEAT utilizes the finite volume discretization solver in OpenFOAM, allowing the use of unstructured meshes and arbitrary geometries. The author elaborated on some of the key physics features that have been incorporated into OFFBEAT that are consistent with other proprietary fuel performance codes. The performance of OFFBEAT has been validated against fuel performance benchmarks available publicly. The code has also been coupled with neutron transport codes. The ongoing developments are focused on expanding the validation of the code, extensions to the physical models for fuel performance and improving the fuel–cladding interaction models.

A–3.4. SCIANTIX open-source code for fission gas behaviour: objectives and foreseen developments

The author gave an overview of the SCIANTIX open source code for fission gas behaviour. The SCIANTIX code was developed to provide a complementary open source code for modelling the physical phenomena of fission gas production and migration in reactor fuels. The SCIANTIX code models both the intragranular and intergranular behaviour of fission gas. The code has been validated against separate effect experiments for intragranular and intergranular bubble swelling, showing good results. Additionally, the code has been coupled with the TRANSURANUS fuel performance code, as well as GERMINAL and OFFBEAT. Future endeavours include hosting the code on a Gitlab repository, development of an object oriented version, improved documentation, standardization of verification and inclusion of regression tests.

The discussions focused on the codes' capabilities, access, ease of use, licensing particularities, documentation, verification, validation and others. The attendees expressed significant interest in each of these codes and in the possibilities of coupling them together into an integrated, multiphysics system for modelling simulation of nuclear reactors.

A–3.5. ONIX: An open-source depletion code for reactor analysis and nuclear security

The author is the main developer of ONIX, an open source tool for depletion analysis. The main aim of the code is to estimate the production of plutonium and tritium in a nuclear facility by coupling ONIX — which is responsible for the depletion calculation — with OpenMC for the transport analysis. Among the key features of ONIX are the ability to find an efficient construction of the burnup matrix, the possible use of precalculated one-group cross-sections with dependence on temperature and composition, and a visualization tool. The plutonium inventory is calculated through the fluence, and it is possible to identify which of the nuclides are most suitable to be employed as fingerprints in terms of plutonium and tritium production control. Given the application, the conclusion and discussion mainly focused on the export control issues and how to avoid barriers in the code development. As a lesson learned, the author suggested involving lawyers from the very beginning to understand if a code can be maintained as open source.

A–3.6. Scalable tightly-coupled multi-physics for fusion reactors with AURORA

The author highlighted the motivations for an open source tool in the nuclear fusion engineering field, pointing out the need for a high fidelity engineering tool for the design of future fusion reactors (the target for a prototype plant in the United Kingdom strategy is 2040). The demanding deadline calls for numerical tools helping the design in terms of geometry creation, engineering simulation, V&V, uncertainty quantification and optimization. The author suggested that the use of prevalidated tools may also accelerate the development process, considering also the complexity of fusion reactor modelling (e.g. tritium transport). Key requirements for a tool are scalability, portability, flexibility, adaptability and maintainability.

In this regard, MOOSE has been selected for the development of the open source framework. The open source code AURORA combines OpenMC for the neutron transport with Direct Accelerated Geometry Monte Carlo (DAGMC) for particle tracking and MOOSE for the heat conduction and thermal expansion. The author also pointed out the features of AURORA in terms of QA compliance and scalability, in addition to the ongoing development of other tools focusing on other physics. The question and answer discussion pointed out the urgency of establishing a good governance model for this and for all open source codes in general.

A–3.7. Retrospect, status and perspective of the development of 'containmentFOAM'

The author shared the motivation for starting the development of containmentFOAM, an open source code based on OpenFOAM for analysing containment pressurization, flows, as well as H_2, CO and aerosol behaviour. In addition, the application range included experimental support, investigation of safety systems, and education and training. The motivations for developing an open source alternative to the legacy proprietary codes are the FAIR principles for publicly funded research as well as the visibility of researchers' work, reusability and facilitation of international collaboration.

The presentation covered the current status of containmentFOAM development in terms of flow and transport phenomena and components. In terms of challenges related to the development of an open source code, the author mentioned the need for coordination, the long term maintenance, the different backgrounds of the users with different application requirements (not always aligned with the actual capabilities of the code) and the complexity of the specific developments. In terms of lessons learned and best practices, the presentation highlighted the importance of proper documentation linked to a GitHub or GitLab repository, done in parallel with code development, early engagement with lawyers for expert control issues and relevant communication channels with effective user support. In terms of user experience, containmentFOAM adopts an ad hoc developed user interface that enables entry for new users and fallback for bug tracking.

A–4. POSTER SESSION

The poster session covered a variety of topics, from open source tools for fusion and codes for corrosion control of fuel cladding to multiphysics coupling techniques for modelling of molten salt reactors and use of open source codes for digital twin development. Table A–1 provides a list of the presented posters.

A–5. PANEL 1: INDUSTRIAL USES OF OPEN-SOURCE NUCLEAR SOFTWARE

The research and education communities are highly engaged in OSS development. However, the power plant vendors who develop power plants and the utilities that own and operate power plants are also users, developers and interested parties in the topics of the IAEA'S Open Source Nuclear Codes for Reactor Analysis (ONCORE) initiative. This panel featured people at the interface between specialized nuclear software and all kinds of user, including many industrial users. The panel consisted of the following participants:

- Moderator: N. Touran worked for 11 years on nuclear methods and analysis at TerraPower on sodium cooled and molten salt reactors and has spent the past two years working on nuclear design configuration management.
- T.E. Valentine is the Director of the Radiation Safety Information Computational Center at Oak Ridge National Laboratory in the United States of America. His research has spanned Monte Carlo code development,

TABLE A–1. PRESENTED POSTERS[a]

Presenter	Country	Title
R. Lewis	United Kingdom	Investigating efficient coupling of Finite Element Analysis with Machine Learning to optimise a laboratory experiment
S.M. Kunju	United Kingdom	Multi-scale computational analysis to predict the irradiation-induced change in engineering properties of fusion reactor materials
A. Davis	United Kingdom	Open source tools in support of multiphysics simulation for fusion
J. Darona	France	Verification & Validation process in the open source TRUST/TrioCFD Platform
J. Groth-Jensen	Denmark	Verification of multiphysics coupling techniques for modelling of molten salt reactors
C. Demazière	Sweden	Active coding assignments in nuclear reactor modelling
R. Lehnigk	Germany	Development and maintenance of OpenFOAM_RCS for German nuclear safety research related to the reactor cooling system
C. Fiorina	Switzerland	The Gen-FOAM multiphysics solver for reactor analysis: status and ongoing developments
J. Stewering	Germany	Validation of an OpenFOAM gas distribution solver for containment analysis
J. Browning	United States of America	Foundations for a fission battery digital twin
J. Romero-Barrientos	Chile	Modification of open source Monte Carlo code OpenMC to include time-dependence in a fissile system including individual delayed neutron precursors
Seoyoon Jeon	Republic of Korea	VANGARD: A GPU-based high-speed pinwise nodal core analysis code
A. Ivanov	Russian Federation	Development of codes for calculating the corrosion of steel cladding of fuel elements with mixed nitride fuel, taking into account the segregation of impurities at grain boundaries

TABLE A–1. PRESENTED POSTERS[a] (cont.)

Presenter	Country	Title
A. Khakim	Indonesia	A concept of fluid dynamic code for molten salt reactor analysis with OpenFOAM
N. Touran	United States of America	An open source data-centric representation of the fast flux test facility isothermal benchmark core
W. Vechgama	Thailand	Development of accessible codes for nuclear safety analysis in Thailand
M. Rocha	Brazil	Development of an open source tool for water hammer and fluid-structure interaction simulation in nuclear reactor systems
A. Purwaningsih	Indonesia	Development of Post Processor for VSOP'94 Output
E. Knudsen	Denmark	Enabling techniques within OpenMC for dynamic models of molten salt reactors
T. Helfer	France	MFront: an open source code generation tool for the rigorous management of material knowledge in the PLEIADES and MAP platforms

[a] The extended abstracts are available on the dedicated Indico site: https://conferences.iaea.org/event/247/

critical and subcritical experiments, cross-section measurements, as well as nuclear and energy policy.

- M. Freitag received a PhD in mechanical engineering from Darmstadt University, Germany, in the field of CFD and modelling of gas turbine combustors. In 2011, he joined Becker Technologies, a private research company most acknowledged for the operation of the THAI facility, which is an experimental test facility investigating design basis accidents and severe accident conditions in nuclear containment. THAI data are used worldwide for validating CFD, and lumped parameter codes, and he oversees the operation of the facility and the organization of international code benchmarks.
- M. Fleming is the Head of the NEA Data Bank. He has a PhD from King's College London in Theoretical and Mathematical Physics and was a section leader at the United Kingdom Atomic Energy Authority (UKAEA) in nuclear data and methods.

The first question focused on the specialized needs of industrial users of nuclear software that are less common in academic or research settings. Industrial users need a good answer quickly, economically and with a rigorous QA pedigree. They are often less interested in cutting-edge computational endeavours than are R&D institutions. It was explained that industrial users require formal documentation and QA records, including a requirements specification, implementation documentation, a user manual, a developer manual, a test verification report and experimental validation reports. It was also mentioned that many open source tools can be published relatively easily compared with commercial packages, and as such there are some open source packages that lack sufficient documentation.

The panel discussed specific QA standards, such as ISO 9001 and ASME NQA-1. There was a significant amount of discussion of MCNP and SCALE. While these are not open source, they are among the Radiation Safety Information Computational Centre's most requested codes from industrial users. It was further mentioned that the open source model is ideal for research and education. It was discussed that industrial users can choose either to procure a qualified code from a qualified vendor or to take an existing code and dedicate it under their own QA programme, becoming the custodians of the existing code themselves. One benefit of open source codes is that installing them on high performance computing systems does not add an institutional burden of obtaining additional licences for system administrators. The audience and panel went back and forth about the burden of obtaining access to controlled codes.

The second question was about the allure of OSS. Industry often asks for dedicated professional support and documentation through a service contract, and switching from an established commercial code to a newer open source one would be a major challenge. Additionally, the total cost is of prime importance to industrial users. While licensing fees are obviously lower for open source options, additional costs come from training, updates, QA services and so on, sometimes leading to higher total costs for industrial users.

It was acknowledged that the RedHat-like model, where a company can provide professional services around OSS, could potentially provide this kind of appeal to industrial users, although the panel and audience did not know of any. RedHat has a much larger potential user base (all businesses worldwide) compared with a nuclear specialist company, so the business case may be more challenging.

The audience asked the moderator what the motivations were for making the ARMI code open source. He expressed the hope that if university students used the AMRI code during education, then they would be able to onboard faster if they came to the company. Additionally, if other users add capabilities to the code, then the company could take advantage of their work. Finally, a big picture motivation was given of helping the community progress efficiently by sharing efforts on the more commodity oriented aspects of modelling and simulations.

A–6. PANEL 2: VISION AND PATH FORWARD FOR OPEN SOURCE SOFTWARE TO SERVE THE NUCLEAR COMMUNITY

The panel consisted of the following participants.

Moderator: B. Forget is Professor of Nuclear Engineering at Korea Electric Power Co (KEPCO) and Department Head of the Nuclear Science and Engineering Department at the Massachusetts Institute of Technology (MIT), United States of America. His research interests are in the areas of computational reactor physics, radiative transport and high performance computing. He is also the founder of the Computational Reactor Physics Group, which performs R&D in advanced deterministic transport methods, advanced stochastic transport methods, algorithms for high performance computing, acceleration/multiscale methods and multiphysics methods.

P. Romano is the main developer of the OpenMC open source Monte Carlo code, which was part of his PhD work at MIT. He is now leading the development of this code from Argonne National Laboratory, supported by various research contracts with the USDOE, Office of Science. The OpenMC community has grown considerably over the past few years, but funding keeps it aligned with research contract priorities. An essential goal is to demonstrate value to sponsors of the open source model and help to grow the code. One key future direction discussed was a further increase of modularity of the software to facilitate contributions and growth in many areas of research.

A. Dufresne is the head of the Computer Program Services division of the NEA Data Bank, which has served as a unique platform for distribution and knowledge transfer within the NEA for more than 60 years. The collection includes benchmarks, safety joint projects, integral experiments and handbooks. The CPS has also increased its contributions to e-learning and has been adding new courses to help to promote the use of OSS, with a recent OpenMC course offered in 2022. A new initiative using GitLab aims to strengthen the connections between the different elements of the CPS and improve delivery of that information. The CPS wants to help open source developers to gain more visibility and develop a stronger user community. One key component of the FAIR principle that was highlighted during the discussion was 'interoperable', and it was noted that much work still remains to be done in the community to achieve this goal.

D. Gaston is a long-time developer of large computational software efforts and has a background in computer science. He started the MOOSE project in 2008 and is the deputy director of the USDOE's Nuclear Energy Advanced

Modeling and Simulation (NEAMS) programme and currently working at Idaho National Laboratory. He is a strong believer in open source projects and has contributed to many different packages over the years. The key to successful projects is the community; this involves quick turnaround, constant and clear information and offering training on a regular basis. Another key component to facilitate sustainability is automation, from testing, code format checking, downstream integration, branch promotion, documentation, web site updates and training material. The future revolves around further growing the community by constantly improving training and support, providing better integration of physics modules, as well as improving testing. However, export control regulations are complex and need to be clarified and addressed.

S. Kelm has been the lead developer of containmentFOAM at Forschungszentrum Jülich and supervises a large research group that is further developing the software. He strongly supports the FAIR principle for publicly funded research. He strongly advocates that open source projects need to maintain documentation in parallel with code development and ought to always consider the users. Developers also need to think of maintainability and code quality. Best practice also includes internal continuous integration and testing. Key challenges that were encountered involved incompatible workflows for users that were already familiar with OpenFOAM but were trying to integrate some components of containmentFOAM. Difficulties are encountered in getting users to use models consistently. A key suggestion for the future is to help developers with ways of growing the community and increasing visibility.

The first question dealt with the lack of common standards or heterogeneous contributions in OSS and with the best way to deal with such issues. There was no agreement within the panel about a set of common standards, since each code was designed at different times to address different issues, and not all codes are necessarily in the same language. However, the importance of clear guidelines and developer documentation was emphasized. Additionally, a clear line of communications can avoid surprise contributions and help with integration, since standards can be applied from the start. The panel also highlighted the importance of automation for format checking, documentation and testing, which truly facilitates many issues that can arise.

The second question focused on export control regulations and the value of these regulations. While this topic elicited a lot of discussion, there was no clear answer. There was agreement that the laws across all countries are difficult to interpret, leading to great confusion. There was a recommendation to consult with legal counsel within institutions before launching any new open source project.

The third question focused on how best to develop a community of developers and users. The discussion centred on modern software resources such as GitHub and GitLab, discussion forums, Slack channels or Discourse channels. It was noted that many of these resources are made freely available to open source projects and facilitate communication and community building. Additionally, one main pathway for increasing usage is through user training. The NEA and the IAEA have been very supportive in organizing training workshops by helping with organization, recruiting and hosting. An additional point was also made about early adoption of a strong governance model: it is preferable to have something in place before an issue appears.

The fourth question focused on a funding model and maintaining support for developers and users. Once a project gains a certain reputation, base funding is provided by certain institutions, but getting there takes time and relies heavily on the generosity of the developers and sponsors. The difficult part is the transition from a single group lead project to a multi-developer, multi-institution project. Heavy reliance on a few developers is unsustainable, and support needs to be obtained by an institution that appreciates the value of open source development. The NEA and the IAEA play a role in helping to develop a user community that can help to demonstrate the value of OSS to key institutions.

A–7. PANEL 3: OPEN-SOURCE SOFTWARE FOR EDUCATION AND TRAINING

Computer modelling and simulation tools have been extensively used in nuclear science and engineering education and training. The participants in this panel presented good examples of using codes in education and training from their respective organizations. Although the meeting was on open source codes, the good examples were not restricted only to OSS. The panel's idea was to inspire the meeting attendees with innovative examples and through interactive discussions to address the following question. Is OSS always the right solution depending on the purpose of the course/training? Possible application areas of OSS for education and training are the following:

— Learning to use a modelling software;
— Investigating physical phenomena using a modelling software;
— Learning modelling algorithms and modelling skills.

The panel consisted of the following participants:
Moderator: C. Demazière is a Professor of Subatomic, High Energy and Plasma Physics at Chalmers University of Technology, Sweden, where he leads the Deterministic REactor Modelling (DREAM) task force. DREAM is a cross-disciplinary group with expertise in neutron transport, fluid dynamics, heat transfer and numerical methods. The group is working on beyond state of

the art techniques for modelling nuclear reactors, thus contributing to improved simulation tools and enhanced safety. He lectures in courses on the physics and modelling of nuclear reactors that deal with the multiphysics and multiscale aspects of such systems. He is a member of the American Nuclear Society. His research interests are in reactor physics, neutron transport, coupled neutronics/thermal hydraulics, neutron noise and reactor kinetics, thermal hydraulics, signal processing and analysis, and uncertainty and sensitivity analysis.

Moderator: K. Ivanov is a Professor of Nuclear Engineering and Head of the Nuclear Engineering Department at North Carolina State University (NCSU), USA. His expertise and experience are in developing methods and computer codes for multidimensional reactor analysis including analysis of normal and abnormal operations. His most recent research concerns reactor simulations and sensitivity feedback, including reactor design and optimization calculations, safety analysis, V&V and uncertainty quantification, propagation and reduction. His expertise is in reactor physics, fuel physics and multiphysics for multidimensional reactor core analysis including novel high fidelity modelling and simulation, high-to-low model fidelity information, data driven methodologies and machine learning. He has led nine international programmes supported by NEA/OECD, the United States Nuclear Regulatory Commission, and USDOE, and is the Chair of the Working Party on Scientific Issues and Uncertainty Analysis of Reactor Systems at the Nuclear Science Committee, OECD/NEA.

M. Avramova is a Professor of Nuclear Engineering, University Faculty Scholar, Director of the Consortium for Nuclear Power, and Coordinator of the COBRA-TF (CTF) User Group at the Nuclear Engineering Department of NCSU, USA. Her expertise and experience include development and V&V of core thermal hydraulics and multiphysics models and codes for reactor design, transient and safety computational analysis. Her latest research efforts have been focused on high fidelity multiphysics simulations, as well as on uncertainty and sensitivity analysis of reactor design and safety calculations. She is a co-chair of the Expert Group on Reactor Core Thermal-Hydraulics and Mechanics, under the guidance of the NEA/OECD Nuclear Science Committee's Working Party on Scientific Issues and Uncertainty Analysis of Reactor Systems. She leads several ongoing OECD/NEA international benchmark activities.

Z. Elter is a neutronics specialist at Seaborg Technologies, Denmark, and a researcher at Uppsala University, Sweden. His research is related to nuclear safeguards (non-destructive assay). He has performed particle transport simulation and burnup calculations of spent nuclear fuel and has utilized machine learning techniques for verification of spent fuel. One of his responsibilities is performing radiation protection and neutron activation studies of the NESSA facility. His teaching experience includes data labs related to nuclear reactor core studies and nuclear physics laboratory exercises.

S. Lorenzi is an Assistant Professor at the Nuclear Reactors Group, Department of Energy, Polytechnic University of Milan, Italy. His research has spanned linear stability analysis and dynamics characterization, object oriented modelling, control strategy investigation, development of multiphysics modelling, lead cooled fast reactors, molten salt reactors, nuclear hybrid systems and experimental activities related to TRIGA reactors.

B. Forget, biography as in previous panel discussion.

The first question focused on the general advantages and disadvantages of using open source codes (as opposed to closed source software). It was emphasized that the use of open source tools can allow the adoption of a problem based learning approach as a teaching method for nuclear knowledge, complementary to traditional learning. It can not only attract the attention and the involvement of the students, but also help to effectively link theoretical concepts and practical experiences, involving them in a recursive process that combines knowledge, competence to solve problems and capacity to analyse results.

It was further discussed how computer codes and software packages and associated databases and data libraries (including open source codes and software and open access data and libraries) are used in education and training (in addition to R&D) at NCSU. Open source software packages such as GeN-Foam, MOOSE, Modelica/Transform, OpenMC and OpenMOC are utilized. There are also in-house open source developments such as an open source, parallel and distributed web based probabilistic risk assessment platform to support real time nuclear power plant risk-informed operational decisions — OpenPRA Initiative as well as the JENES project, which aims to enhance daily nuclear engineering simulations and computations by using the Julia language for different nuclear applications. The advantages of open source codes are that modifications and improvements can be included as part of the education and training process.

The conclusion is that students enjoy active learning (research-like) and open teaching material reaches more people. An advantage of OSS is that it facilitates export control negotiations for research contracts.

The second question touched on the pitfalls of using OSS for education and training specifically. The four panellists iterated some of the pitfalls already mentioned in other panels: maintenance, QA process, poor documentation, lack of revision control and formal V&V procedures, compliance with ISO 9001 and/or NQA-1 standards, and unclear copyright and licence processes and associated contributor licence agreements. These pitfalls can affect the trustworthiness and reliability of using OSS for education and training.

The third question dealt with ways in which OSS better supports students in their learning. The following ways were listed: it simplifies the use of numerical tools, eliminating the main issues of intellectual property and licensing; it lowers the entry barrier for nuclear newcomer countries, thus promoting international collaboration; it facilitates the sharing and the reuse of solvers, routines and exercises among different institutions and among students, without the need to start a model from the beginning, but instead focusing on the specific topic of a module or lecture (this is important in the case of short courses with a limited amount of time); it motivates students and young researchers to implement, test and promote their development ideas, since they can also rely on a typically active support community of developers and users; it makes possible the implementation of an efficient problem based learning approach, specifically devising learning activities aligned with the intended learning outcomes; it permits one to explore the code structure and to 'change' the physics for a more comprehensive understanding of both the theory and the software; it increases the students' skills with tools that are increasingly being used in the working field; and it provides widely applicable and reusable knowledge, not restricted to codes that can be used only in specific institutions.

Open source also allows guided exercises, which are appreciated by students. This fact stimulates a shift to the Python language, for example extending Python use for a full reactor physics introductory course for distance and active learning.

It was proposed to introduce special academic courses on OSS addressing issues such as code version control using GitHub or GitLab, coding languages, compiling, code performance and optimization, and several other topics. Computational skills are very important for nuclear science and engineering students, and it is important and attractive to students to be able to show in their CVs such skills, through courses, certificates and minors in computational science or computational physics.

The last question focused on examples of 'software community building' resulting from courses or training on OSS. All four panellists emphasized that open source tools have been already adopted in universities and research centres for education and training efforts. Software community building has been performed in the form of user groups around the OpenFOAM toolkit, OpenMC and OpenMOC codes' activities. The Polytechnic University of Milan has organized an 'OPENER' summer school for international PhD students, where the use of open source tools allowed students to be guided through the development of a simple but complete multiphysics solver for the analysis of representative nuclear engineering problems, thus enhancing their understanding of the underlying physics.

It was noted that the current education activities of the OECD/NEA Nuclear Science Committee include developing open access benchmarks for education by the Working Party on Scientific Issues and Uncertainty Analysis of Reactor Systems (WPRS). WPRS produced comprehensive sets of benchmarks in reactor physics, thermal hydraulics, multiphysics, and shielding modelling and simulation. For these benchmarks, computer codes or software from the NEA Data Bank and/or open source codes or software and open access data or libraries can be used. The WPRS is preparing benchmarks as 'student training sessions', which can be used in classroom sessions or as remote classes on the NEA Learning Management System (LMS) platform, as follows:

— Existing specifications serve as a basis for exercises.
— Exercises are to be based on sample problems:
  • Input deck sets are provided by benchmark participants which include 'holes' (i.e. need to be adjusted/adapted);
  • Results are to be extracted and interpreted by students.
— Docker environments with all required tools are provided by the NEA Data Bank; the software tool chain needs to be open source or accessible via the NEA Data Bank.
— Supplemental lectures on reactor technologies are included.
— Final sessions with student presentations and expert attendance are carried out.

The proposed delivery format is webinars supported by the NEA LMS platform.

ABBREVIATIONS

API	application programming interface
ARMI	Advanced Reactor Modeling Interface
ASME	American Society of Mechanical Engineers
BSD	Berkeley source distribution
CAD	computer aided design
CANDU	Canada deuterium–uranium (reactor)
CFD	computational fluid dynamics
CPS	Computer Program Services
CPU	central processing unit
CSS	closed source software
DOI	digital object identifier
ENDF	Evaluated Nuclear Data File (library)
EPFL	École Polytechnique Fédérale de Lausanne
FAIR	findable, accessible, interoperable and reusable
FEM	finite element method
FVM	finite volume method
GPL	(GNU) General Public Licence
GPU	graphic processing unit
HCP	HTR Code Package
HPC	high performance computing
IT	information technology
JEFF	Joint Evaluated Fission and Fusion (library)
LFMSR	liquid fuel molten salt reactor
LGPL	(GNU) Lesser General Public Licence
MCNP	Monte Carlo N-Particle Transport Code
MOOC	Massive Open Online Course
MOOSE	Multiphysics Object Oriented Simulation Environment
NEA	Nuclear Energy Agency
NEORL	NeuroEvolution Optimization with Reinforcement Learning
NQA	Nuclear Quality Assurance
NSG	Nuclear Suppliers Group
OECD	Organisation for Economic Co-operation and Development
OFFBEAT	OpenFOAM Fuel BEhavior Analysis Tool
ONCORE	Open source Nuclear Codes for Reactor Analysis
OSS	open source software
PWR	pressurized water reactor

QA	quality assurance
Q&A	question and answer
R&D	research and development
USDOE	United States Department of Energy
V&V	verification and validation

CONTRIBUTORS TO DRAFTING AND REVIEW

Batra, C.	International Atomic Energy Agency
Brooks, H.	United Kingdom Atomic Energy Authority, United Kingdom
Choe, J.	Korea Atomic Energy Research Institute, Republic of Korea
Davis, A.	United Kingdom Atomic Energy Authority, United Kingdom
Demazière, C.	Chalmers University of Technology, Sweden
Fiorina, C.	Texas A&M University, United States of America
Forget, B.	Massachusetts Institute of Technology, United States of America
Gladyshev, M.	International Atomic Energy Agency
Hébert, A.	Polytechnique Montréal, Canada
Ivanov, K.	North Carolina State University, United States of America
Kelm, S.	Forschungszentrum Jülich, Germany
Kriventsev, V.	International Atomic Energy Agency
Liu, X.	International Atomic Energy Agency
Lorenzi, S.	Polytechnic University of Milan, Italy
Morelová, N.	International Atomic Energy Agency
Munk, M.	University of Illinois Urbana-Champaign, United States of America
Romano, P.	Argonne National Laboratory, United States of America
Scolaro, A.	École Polytechnique Fédérale de Lausanne, Switzerland
Shriwise, P.	Argonne National Laboratory, United States of America

Shwageraus, E.	University of Cambridge, United Kingdom
Touran, N.	TerraPower LLC, United States of America
Valentine, T.	Oak Ridge National Laboratory, United States of America
Virgili, N.	International Atomic Energy Agency
Wilson, P.	University of Wisconsin–Madison, United States of America
Yang, X.	International Atomic Energy Agency

Technical Meeting

Milan, Italy: 20–24 June 2022

Consultants Meetings

Vienna, Austria: 9–10 February 2021, 6–7 July 2021,
15–16 March 2022, 21 September 2022, 27 March 2023

CONTACT IAEA PUBLISHING

Feedback on IAEA publications may be given via the on-line form available at:
www.iaea.org/publications/feedback

This form may also be used to report safety issues or environmental queries concerning IAEA publications.

Alternatively, contact IAEA Publishing:

Publishing Section
International Atomic Energy Agency
Vienna International Centre, PO Box 100, 1400 Vienna, Austria
Telephone: +43 1 2600 22529 or 22530
Email: sales.publications@iaea.org
www.iaea.org/publications

Priced and unpriced IAEA publications may be ordered directly from the IAEA.

ORDERING LOCALLY

Priced IAEA publications may be purchased from regional distributors and from major local booksellers.

Printed and bound by CPI Group (UK) Ltd, Croydon, CR0 4YY

06/07/2026

02160596-0002